# Smart and Advanced Ceramics and Applications

*Edited by Mohsen Mhadhbi*

Published in London, United Kingdom

Smart and Advanced Ceramics and Applications
http://dx.doi.org/10.5772/intechopen.97891
Edited by Mohsen Mhadhbi

Contributors
Stevan Ljubinko Stupar, Amparo Borrell, Rut Benavente, Maria Dolores Salvador, Sheela Devi, Venus Dillu, Mekonnen Tefera Kebede, Anil Kumar Singh, Rasmita Jena, Kouru Chandrakanta, Marina V. Kalinina, Daria A. Dyuskina, Sergey V. Myakin, Irina G. Polyakova, Olga A. Shilova, Maxim Yu. Arsent'ev, Mohsen Mhadhbi, Ali Sohani



First published in London, United Kingdom, 2022 by IntechOpen
IntechOpen is the global imprint of INTECHOPEN LIMITED, registered in England and Wales, registration number: 11086078, 5 Princes Gate Court, London, SW7 2QJ, United Kingdom

British Library Cataloguing-in-Publication Data
A catalogue record for this book is available from the British Library

Additional hard and PDF copies can be obtained from orders@intechopen.com

Smart and Advanced Ceramics and Applications
Edited by Mohsen Mhadhbi
p. cm.
Print ISBN 978-1-80355-864-6
Online ISBN 978-1-80355-865-3
eBook (PDF) ISBN 978-1-80355-866-0

# Meet the editor

Dr. Mohsen Mhadhbi obtained his Ph.D. from the Faculty of Sciences of Sfax, Tunisia. He is currently an Assistant Professor of Chemistry at the National Institute of Research and Physical-Chemical Analysis, Tunisia. His research interests include nanomaterials, powder technology, ceramics, modeling, and simulations. Dr. Mhadhbi has published works in national and international journals and books. He is a member of various scientific journals and associations and has been an editorial board member for several journals.

# Contents

**Preface** **IX**

**Section 1**
Introduction to Smart and Advanced Ceramic Materials **1**

**Chapter 1** **3**
Introductory Chapter: Smart and Advanced Ceramics and Applications
*by Mohsen Mhadhbi and Ali Sohani*

**Section 2**
Ceramics Based Nanomaterials **7**

**Chapter 2** **9**
Synthesis and Characterization of NanoBismuth Ferrites Ceramics
*by Sheela Devi, Venus Dillu and Mekonnen Tefera Kebede*

**Chapter 3** **33**
Ballistic Composites, the Present and the Future
*by Stevan Stupar*

**Section 3**
Refractory Ceramics **51**

**Chapter 4** **53**
Structural, Magnetic, and Magnetodielectric Properties of Bi-Based
Modified Ceramic Composites
*by Rasmita Jena, Kouru Chandrakanta and Anil Kumar Singh*

**Section 4**
Smart Ceramic Materials **69**

**Chapter 5** **71**
Design and Development of Zirconia-Alumina Bioceramics Obtained
at Low Temperature through Eco-Friendly Technology
*by Rut Benavente, Maria Dolores Salvador and Amparo Borrell*

**Chapter 6** 87
Synthesis and Investigation of Ceramic Materials for Medium-Temperature Solid Oxide Fuel Cells
*by Marina V. Kalinina, Daria A. Dyuskina, Irina G. Polyakova, Sergey V. Mjakin, Maxim Yu. Arsent'ev and Olga A. Shilova*

# Preface

Ceramic materials have attracted the attention of scientists, researchers, and engineers due to their potential to manipulate the length scale in nanometers. These materials possess excellent properties, which give them a variety of industrial uses, including in the automotive, aerospace, and biomedical industries.

The book *Smart and Advanced Ceramic Materials and Applications* provides a comprehensive overview of smart and advanced ceramic materials and their applications. The first part of the book focuses on both ceramics-based nanomaterials and ceramic sensors. The second part of the book provides an overview of smart ceramic materials, which are ceramic materials prepared from ultrafine particles. These materials play a crucial role in challenging applications such as biomedical and dental applications. The final section of the book discusses refractory ceramics.

The book will benefit students, scientists, researchers, engineers, and academics interested in traditional, smart, and advanced ceramic materials.

I wish to express my gratitude to the contributing authors for their excellent chapters. This book could not have been successfully published without the hard work of the authors as well as the staff at IntechOpen.

**Mohsen Mhadhbi**
National Institute of Research and Physicochemical Analysis,
Ariana, Tunisia

Section 1

# Introduction to Smart and Advanced Ceramic Materials

Chapter 1

# Introductory Chapter: Smart and Advanced Ceramics and Applications

*Mohsen Mhadhbi and Ali Sohani*

## 1. Introduction

Ceramic materials are inorganic compounds consisting of metallic and non-metallic elements, which are held together with ionic, and/or covalent bonds [1]. It combines the principles of physics, chemistry, and engineering [2]. However, smart and advanced ceramic materials are usually fabricated in small quantities and at high costs. Furthermore, these new class of materials play a major role in the high-tech industry, energy, biomedical, military industry, solar cells, and fuel cells due to their specific high-temperature mechanical and optical properties, their biocompatibility, and their unique composite effects of light, sound, electricity, magnetism, heat, or function. In this context, many excellent researches have been reported on smart and advanced ceramic materials [3–8].

Various types of smart and advanced ceramic materials each have particular properties, providing high performance and economical replacement to traditional materials, such as glass, plastics, metals, and others. Indeed, advanced ceramics applications attempt to improve operation at low cost and to address the requirements of the particular applications.

However, the widely used smart and advanced ceramic materials are shown in **Figure 1** and developed below:

- *Alumina*: Alumina ($Al_2O_3$) is one of the most widely used advanced ceramic materials. It has excellent properties, such as high hardness, high melting point, high mechanical strength, excellent wear resistance, high corrosion resistance, low density, etc. This material is a kind of ceramic with a wide range of uses.

- *Zirconia*: Zirconia ($ZrO_2$) possesses desirable properties, such as low thermal conductivity, high mechanical strength, excellent wear resistance, high corrosion resistance, and good biocompatibility, making it one of the most popularly utilized advanced ceramic materials.

- *Cubic boron nitride*: Cubic boron nitride (CBN) is a most important advanced ceramic material, which is prepared through compressing hexagonal boron nitride at high temperature and high pressure. It possesses excellent properties including high hardness, high chemical resistance, high thermal stability, high electrical resistivity, high thermal conductivity, etc.

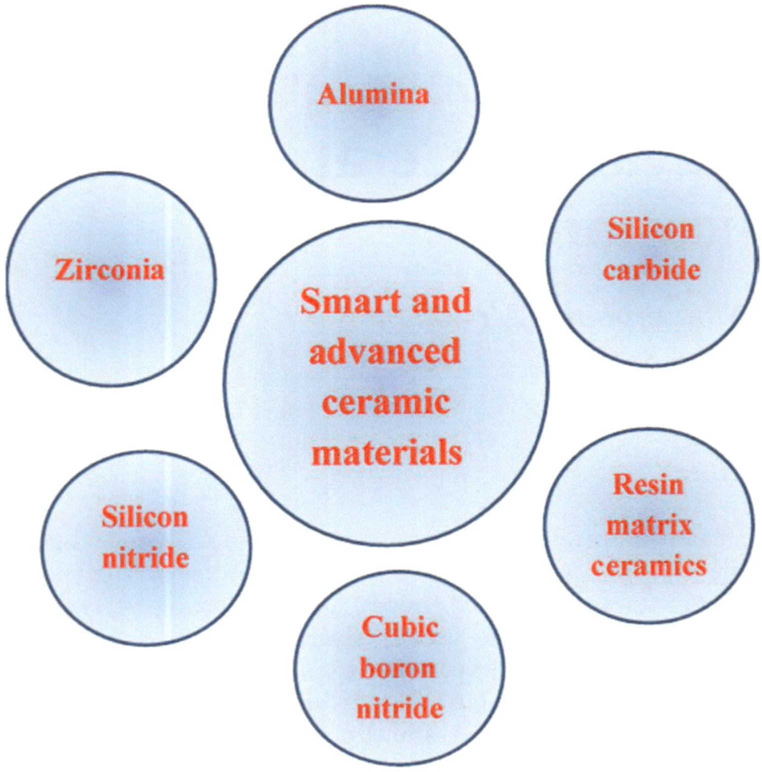


**Figure 1.**
*Different types of smart and advanced ceramic materials.*

- *Silicon nitride*: Silicon nitride ($Si_3N_4$) is one of the few advanced ceramic materials that is able to survive the severe thermal gradients and thermal chock stresses. This material is widely used in structural applications due to its specific properties, such as low density, good wear resistance, excellent oxidation resistance, etc.

- *Silicon carbide*: Silicon carbide (SiC) is a particular advanced ceramic material, which is largely used as abrasive and cutting tools because of its unique characteristics, such as low density, great hardness, and strength, high thermal stability, and excellent corrosion resistance, etc.

- *Resin matrix ceramics*: Resin matrix ceramics (RMCs) is a novel class of advanced ceramic materials which combines the properties of polymer and ceramic matrices. There are two types: resin-based ceramics and hybrid ceramics, which are principally applied in dentistry.

Smart and advanced ceramic materials are referred to as special and engineering ceramics. **Table 1** shows the various applications of advanced ceramics which include two types.

Smart and advanced ceramic materials provide products and services to an increasing and important range of advanced manufacturing industries, involving automotive, aerospace, energy, defense, telecommunications, etc. More extensive research on the processes of ceramic materials is necessary and extensive

| Structural ceramics | Functional ceramics |
|---|---|
| Bioceramics | Piezoceramics |
| Automotive ceramics | Magnetoceramics |
| Nuclear ceramics | Package ceramics |
| Ceramics used in cutting tools | Optical ceramics |
| Tribiological ceramics | Conductive ceramics |

**Table 1.**
*Applications of smart and advanced ceramic materials.*

development plans should be presented. Thus, the advanced ceramic materials possess some challenges including the high cost of product research and development, the innovation at the industrial scale could acquire a remarkable investment in capital and time, the commercialization of innovation is slow, and the deficiency of cross-sector technology transfer.

## Acknowledgements

The authors gratefully acknowledge the author service manager Ms. Elena Vracaric.

## Conflict of interest

The authors have no conflict of interest to declare.

## Author details

Mohsen Mhadhbi[1*] and Ali Sohani[2]

1 Laboratory of Useful Materials, National Institute of Research and Physicochemical Analysis, Ariana, Tunisia

2 Faculty of Mechanical Engineering-Energy Division, K.N. Toosi University of Technology, Tehran, Iran

*Address all correspondence to: mhadhbi_mohsen@yahoo.fr

## References

[1] Fundamentals of Ceramics [Internet]. 2018. Available from: https://www.brainkart.com/article/Fundamentals-of-Ceramics_6367/

[2] Shree V. What is Ceramic Engineering? Scope and Career Opportunities [Internet]. 2015. Available from: https://www.careerindia.com/courses/unique-courses/what-is-ceramic-engineering-scope-and-career-opportunities-015166.html

[3] Mhadhbi M. Book of Recent Advances in Ceramic Materials for Dentistry. London: Intechopen; 2021. p. 296. DOI: 10.5772/intechopen.87703

[4] Sänger JC, Pauw BR, Sturm H, Günster J. First time additively manufactured advanced ceramics by using two-photon polymerization for powder processing. Open Ceramics. 2020;**4**:100040. DOI: 10.1016/j.oceram.2020.100040

[5] Michaelis A, Scheithauer U, Moritz T, Weingarten S, Abel J, Schwarzer E, et al. Advanced manufacturing for advanced ceramics. Procedia CIRP. 2020;**95**:18-22. DOI: 10.1016/j.procir.2020.11.002

[6] Dey A, Rangappa D, Madhusoodana CD. Advanced ceramics and nanomaterials for sustainable development. Ceramics International. 2019;**45**:24955-24956. DOI: 10.1016/j.ceramint.2019.09.253

[7] Lakhdar Y, Tuck C, Binner J, Terry A, Goodridge R. Additive manufacturing of advanced ceramic materials. Progress in Materials Science. 2021;**116**:100736. DOI: 10.1016/j.pmatsci.2020.100736

[8] Elwaseef MM. Smart materials. Biomaterials Journal. 2022;**1**:11-18. DOI: 10.5281/znodo.5829408

Section 2

# Ceramics Based Nanomaterials

Chapter 2

# Synthesis and Characterization of NanoBismuth Ferrites Ceramics

*Sheela Devi, Venus Dillu and Mekonnen Tefera Kebede*


## Abstract

Multiferroic nanomaterials bear draw attention plenty consideration on account of the mixture of two or more determinants, in the way that ferroelectricity, ferromagnetism, and ferroelasticity, giving an expansive range of professional, depressed capacity, environmentally intimate request. Nano-bismuth ferrite ($BiFeO_3$, BFO) exhibits two together (anti) ferromagnetic and ferroelectric real estate at room temperature. Therefore, it bears risk a very influential part fashionable the multiferroic foundation. This review focuses ahead of the progress of nano-BFO objects, containing unification, facial characteristics, structures, and potential uses of multiferroic order accompanying novel functions. Hopes and danger happen all investigated and made clear. We hope that this review will be a part of a review and encourage more research workers to win accompanying nano-BFO results.


**Keywords:** Nano bismuth ferrites ceramics, properties, characterization, multifunctional device

## 1. Introduction

Nanomaterials are materials with length scales ranging from 1 to 100 nm. The properties of materials change dramatically on the nanoscale when compared to their bulk counterparts. The physiochemical properties of these materials are very sensitive to size and shape, resulting in properties that are completely different when studied in bulk materials. Because of their high surface-to-volume ratio, nanomaterials have a distinct set of properties. At both normal and high temperatures, nano-ceramic materials exhibit excellent refractory properties, chemical resistance, mechanical resistance, and hardness [1] $BiFeO_3$, BFO is an exciting one-phase application that has focused on multidisciplinary device applications due to its unique properties, including Currie high temperature of ferroelectricity (TC = 1103 k), high Neel temperature for antiferromagnetism (TN = 643 k), lead-free piezoelectricity, and impressive photoelectric performance in the visible area [1, 2]. These features make $BiFeO_3$ a promising candidate, especially in ferroelectrics, magnetic, piezoelectrics, Photovoltaic devices, and photocatalytic function. In addition, the combination of these structures has the potential to provide the next generation of electrical appliances with a wide range of functions. The $BiFeO_3$ investigation began in the 1950s. In the Ramesh group (24) in 2003, the discovery of a large polarization of fossils, 15 times larger than previously obtained by large samples (6.1 $\mu C/cm^2$), combined with the strong ferromagnetism seen in small $BiFeO_3$ films, and increased studies. In this field and several other investigations in bulk, a thin film and forms with a $BiFeO_3$ structure have been made since then. Although BFO has a spiral spin formation with a periodicity of 62 nm as a single type of G-type antiferromagnetic material, this combination of the soft magnetoelectric liner makes it difficult to use for many functional devices [3]. These days, it has been found that BFO nanoparticles exhibit strong ferromagnetism due to their magnitude of less than 62 nm [4]. As a result, the successful development of magnetoelectric integration in nanosized BFO has played a significant role in microelectronic devices [5–7]. The importance of the semiconductor circuit combined science and technology have resulted in the feature size of microelectronic devices being reduced to nanoscale magnitude. BFO material features nanoscale novels different from most of their counterparts and films. Knowing the impact of BFO on the nanoscale is essential in promoting the development of new electronic devices. The controlled integration of nano-BFO nanostructures such as nanowires, nanotubes, nano-powders, and arrays has made great effort due to size-dependent effects in structure and limited size. Some gains have been made in classifying buildings. In addition, BFO nanostructures are gaining considerable attention in heterostructures and domain classification [8, 9]. This paper provides an overview of recent developments in the integration, segregation of actors, visual structures, and the potential use of the nanosized BFO. Other issues that need to be addressed in future research are also highlighted.

## 2. Nanoparticles

Nanoparticles are particles with dimensions ranging from 1 to 100 nm that have features not discovered in the bulk material of the likely structure [10].

### 2.1 Synthesis methods

BFO nanoparticles Synthesis is one of the fascinating areas of study because of its technological relevance [11]. Due to the extensive development of nano-BFOs, further research is needed to determine the best ways to make high-quality BFO nanoparticles in purity, shape, size, size distribution, stability, and particle morphology [12]. Currently, the synthesis of BFO nanoparticles has been reported using several methods such as solid-state [13], ferrite precursor [14], hydrothermal [15], sol-gel [16], and co-precipitation [17]. To date, however, there is no established method for nano BFO synthesis, and all of these synthetic methods have their advantages and disadvantages. To date, two different methods have been developed to integrate nano-BFO; another way of going down, which means to combine nanoparticles from atoms or molecules by assembling smaller ones into larger ones; another way of going up and down, which means splitting, recording, or filming a thin film or bulk material into nano sizes like significant cuts into smaller ones. The "bottom-up "method has several integration techniques: co-precipitation, thermal decomposition, hydrothermal, sol-gel, solvothermal, flame spray pyrolysis, sonochemical, vapor deposition, and microwave-assisted microemulsion, and polyol routes, the first four of which are very crowded. Although mechanical grinding and pulsed laser ablation are the most widely accepted "top-down" methods combined. Among these synthetic methods, modified polyol, sonochemical, and microwave irradiation are the most widely used methods in the modern era in synthesizing a variety of nano BFO. These integration methods have several advantages: short reaction time, soft reaction conditions, high yield, and improved selected and clean reaction compared to the other combination methods mentioned above [18]. Wu et al. so far have reported a few of these integration lines [19]. Many of these artificial lines are easily found in books; therefore, in this section, we introduce the most widely used nano BFO integration lines, address their advantages, barrier differences, and summarize current developments.

#### *2.1.1 Sol: gel*

The Sol-Gel method is an easy, popular, and kind of way to go up and down to prepare the nanoparticle. In some cases, modification of precursors usually requires metal salts or metal alkoxide solutions in nanostructures that undergo hydrolysis and condensation polymerization processes to form gels. After being released, the cells turn the jelly cells into something intended to be suppressed. Important processing features of the sol-gel process are the amount of water in the solution, the pH value of the solution, and its temperature. Blending of BFO nanoparticles by sol-gel process, Bi $(NO_3)_3.5H_2O$ and Fe $(NO_3)_3.9H_2O$ are widely used as raw materials [3, 20, 21]. As mentioned below, there are three main sol-gel processes for synthesizing BFO nanoparticle materials.

##### *2.1.1.1 Traditional technique*

The most straightforward process begins by combining iron nitrate with natural solvent (2-MOE or EG) to produce a precisely reddish-black solution and stir at 70-80°C in a hot plate for several hours to obtain a soft gel. Then dry the soft gel obtained by heating to a high temperature (120-160°C) [22–25].

##### *2.1.1.2 Acid assisted sol: Gel technique*

Utilizing this process, acid is added to the solution structure. Tartaric acid $(C_4H_6O_6)$ is a commonly used acid synthesizing BFO nanoparticles [26–29]. By combining iron nitrate with natural solvents (2-MOE or EG), followed by tartaric acid (1: 1 molar ratio iron nitrate), raw material is synthesized using a tartaric acid-assisted sol-gel method. With constant stirring at 60-80°C, until the transparent sol transforms into a brown gel, other recruited acetic acids [23, 30] could also be used to make BFO nanoparticles. The preliminary preparation is similar to the tartaric acid sol-gel method in that the first precursor is made by combining iron nitrate with an organic solvent, but this time the solvent is NH3.H2O before adding acetic acid to the precursor, it is added to adjust the pH of the solution to. The solution is then shaken and stirred at 70°C in a hot magnetic plate to form a sol. The brown gel should then be dried at around 120°C for xerogel powder. In the precursor process, acetic acid acts as an unidenttate-binding agent, and organic solvent (ethylene glycol) acts as a polar-soluble solvent. During heating, Fe ions are octahedrally mixed with condensate acetic ions after adding acetic acid and ethylene glycol. At temperatures of 70°C, this is accompanied by acetate ligands and ethylene glycol. At 120°C, +the resulting esters initiate the formation of Fe-O-Bi bonds [30].

##### *2.1.1.3 The aqueous-based sol: Gel technique*

The latter method is a water-based route where no natural solvent is involved in the precursor mixing. The metal nitrate precursor system dissolves in dilute nitric acid, followed by tartaric or acetic acid [31–34]. Using this line, the metal complex was created by adding acid to the aqueous metal nitrate precursor, and BFO nanoparticles began to form in the calculation after decomposition. Then after any of the routes as mentioned above, the dried gel powder is ground and heated over 400°C for pyrolysis to remove organic impurities and additional calcining around 500-600°C to obtain final nanoparticles [24, 35, 36]. In one study, after final immersion, the samples were repeatedly washed with distilled water, filtered, and dried at 80°C [36]. The sintering temperature is considered an appropriate strategy to monitor the size of BFO nanoparticles. Nanoparticle sizes can vary from less than 15 nm to more than 100 nm by incorporating sintering as low as 350–650°C [31, 34, 37]. In terms of lines (ii and iii), it proves that chelating acids (tartaric acid and citric acid) play an essential role in synthesizing nanoparticles, phases, and morphology. Using a precursor in water, it has been shown that the precipitation temperature of citric acid is less than 100oC than that of the preceding tartaric acid, taking temperatures into BFO nanoparticles that use as much citric acid as possible 350°C [34].

On the other hand, when citric acid was used as an irrigation agent and the calculation temperature was set at 600°C, it was also found that impurities such as $Bi_2O_3$ and $Bi_2Fe_4O_9$ form BFO nanoparticles. On the other hand, the chelating agent of ethylene diamine tetraacetic acid (EDTA) in an aqueous precursor contributes to creating pure-phase BFO nanoparticles called the generation of heterometallic polynuclear complexes in solution [38, 39]. The addition of acrylamide and bisacrylamide monomers has also been beneficial in controlling the specific BFO size by providing a framework for the growth of nanoparticles and adjusting the size of the gel pores, respectively. However, the overgrowth of bisacrylamide may result from particles with no homogeneities and irregular shapes and contaminants [39]. The contaminants were obtained from BFO nanoparticles found in natural solvents (method (i) above) using citric acid as a chelating agent. This is due to a diametric of the citric complex and additional carbonaceous substances, which are pollutants usually created at high temperatures during the automatic combustion process [35, 40]. In comparison, tartaric acid creates bonds and iron ions with two groups of carboxyl and hydroxyl that contribute to a stable polynuclear complex. Esterification gel between metal complex and ethylene glycol [40]. The addition of acid chelating agents often influences the size of the nanoparticles. It is reported that the particle sizes of BFO nanoparticles found in citric acid at 350°C and tartaric acid at 450°C are as small as 4 nm and 12 nm, respectively [34]. The advantages of sol-gel synthesis techniques are easily accessible, energy saving, cost, and performance at low temperatures. The soluble temperature range in the sol-gel range is between 25°C and 200°C, and it is possible to combine a nano BFO with smaller size distribution and a controlled environment. These benefits and their flexibility in nano BFO synthesis make the sol-gel method very attractive. Other advantages of the sol-gel process include the necessary reagents that are simple compounds, producing nanoparticles, no special equipment required, dopants can be easily incorporated into the final product, there is little chance of assembling the particles, and the same grain structure. In addition, it is one of the most popular composite methods for controlling the formation of nanoparticles, microstructure, purity, and stability by adjusting various parameters such as sol concentration, vibration rate, and annealing temperature [41]. In addition, the combination of nanoparticles, films, and coating is the first commercial element of the sol-gel process [16]. The sol-gel synthesis method can synthesize various nanoparticles at a specified temperature range.

### *2.1.2 Hydrothermal*

Hydrothermal is the most common chemical route to nanostructures and particularly nanoparticles metal oxide synthesis. Related to the sol-gel method, it begins with the preparation of a precursor. However, the crystallization phase occurs throughout a hydrothermal process instead of a high-temperature annealing treatment. Final nanoparticles are formed after washing, filtering, and final drying processes.

#### *2.1.2.1 Precursor preparation*

The most popular hydrothermal synthesis process used for the preparation of BFO nanoparticles is the mineralizer-assisted route, in which mineralizers such as KOH (NaOH) [7, 19, 42, 43] or $KNO_3$ [19, 44] are used for the preparation of precursors. This process allows the precursor to be conveniently prepared before hydrothermal synthesis by resolving the metal nitrates in purified water and KOH solution [45]. The aqueous precursor is produced by metal nitrates immersed in dilute nitric acid more generally than others. The method is then applied to the KOH solution for precipitation of Fe3+ and $Bi^{3+}$. Precipitates are purified and cleaned to extract $NO^{3-}$ and $K^+$ ions by pure water. The washed precipitates are then combined with KOH or NaOH solutions and additional $KNO_3$ under continuous agitation [19, 44, 46]. In addition to these processes, the precursor scheme for the alteration of nanoparticles can often involve other additives, such as triethanolamine (TEA) [43] or sodium lauryl sulfate (SDS). In specific, the addition of TEA is observed to create a Fe-TEA complex that inhibits Fe hydroxide precipitation and may minimize synthesis to a maximum of 130oC [43]. Without the presence of the mineralizer, hydrothermal approaches may also be modified [45]. In hydrothermal synthesis of $BiFeO_3$, the precursor of $BiFeO_3$ is frequently synthesized by dissolving the metal sources, Bi and Fe nitrates ($Bi(NO_3)_3.5H_2O$ and $Fe(NO_3)_3.9H_2O$, in deionized (DI) water or citric acid. The metal and citric acid molar ratio is equal to 1:1 ratio. Then, $NH_3$ solution is added to neutralize the unreacted citric acid and change the pH to~9 [45]. Using this updated mineralizer-free method, the dried powder is eventually calcined at 350°C for 6 hr. to extract the final nanoparticles, which is an extremely phase-pure BFO with an average particle size of 55 nm.

#### *2.1.2.2 Hydrothermal reaction*

The solution would be placed in a hydrothermal Teflon linear autoclave at 150°C-220°C for 5–20 hours [19, 47] until the precursor was collected. The autoclave would then be naturally cooled to room temperature, the precipitate collected filtered, and cleaned with deionized water (and 10% acetic acid) to eliminate any soluble salts. Finally, the nanoparticles are obtained by drying the wet samples at 70-160°C for a few hours [19, 44]. A dissolution-crystallization process is assumed to be a synthesis of the BFO Phase by the hydrothermal method. First, $Bi^{3+}$ and $Fe^{3+}$ ions are converted into Fe $(OH)_3$ and Bi $(OH)_3$ hydroxides and later, under hydrothermal conditions, dissolved in a precursor of alkaline mineralizer (KOH, NaOH, LiOH). When the concentration of ions in the alkaline solution reaches the saturation stage, the BFO phase begins to nucleate, followed by crystal formation, and precipitates the supersaturated hydrothermal fluid [19, 48, 49]. BFO particle size and morphology rely on nucleation and crystal growth rates, which are in turn influenced by the degree of supersaturation. The subsequent strong nucleation but low growth rate and smaller particle size would be a high supersaturated precursor with high KOH content and high pH value [42, 49]. The substance's phase and morphologies are greatly influenced by the addition of water during the hydrothermal reaction Han et al. [49] and Chen et al. [19]. Analysis of the BFO hydrothermal synthesis method for the KOH concentration mineralizer, the reaction temperature, and the reaction time was performed. The precursors are held at175-225°C and 200-220°C respectively, for six hours, and produce phase-pure BFO powders using 8 M and 4 M KOH, respectively. Different raw materials and different hydrothermal equipment and conditions can contribute to the observed deviation from nominally optimal conditions (i.e., KOH concentrations). Han noticed that minor $\alpha$-$Bi_2O_3$ could be generated while using a 4 M KOH solution. Scientists have noticed that metal nitrates, under some circumstances, combine with hydroxide ions to form the iron hydroxide phase.

$$2Bi\,(NO_3)_3 + 6KOH \rightarrow Bi_2O_3 + 6\,K^+ + 6NO^{3-} + 3H_2O \tag{1}$$

$$Fe\,(NO_3)_3 + 3KOH \rightarrow Fe\,(OH)_3 + 3\,K^+ + 3NO_3 \tag{2}$$

At a lower concentration of potassium hydroxide, the alpha-$Bi_2O_3$ phase can stay in the precursor, although it can be reduced into Bi3+ at a higher concentration of potassium hydroxide:

$$BiO_3 + 2Fe\,(OH)_3 \rightarrow 2BiFeO_3 + 3H_2O \tag{3}$$

Reaction temperature has a significant effect on the reaction. At lower temperatures, alpha-$Bi_2O_3$ is produced as the key product. By raising the temperature of the solution from 150–175°C, the dissolution of alpha-$Bi_2O_3$ is significantly improved. The BFO process stays the same at the temperature of 225°C [49]. A related result was stated by Chen et al. [19]. Although increasing the pH, reaction time, and temperature will improve the solubility of alpha-$Bi_2O_3$, leading to pure BFO phases. Secondary phases such as the secondary precipitate of excess potassium hydroxide precipitation at a shift in acidic pH may occur when excessive reaction time is at high pH. As OH- ion concentration is becoming higher, particle agglomeration can be expected [19, 49]. Yang et al. figure out the second phase of $Bi_2Fe_4O_9$ forms at a shallow concentration of KOH. Since $Bi^{3+}$ has a greater affinity with OH than $Fe^{3+}$, a low OH-concentration in the hydrothermal reaction solution would lead to more $Fe^{3+}$ ions being dehydrated $Bi_2Fe_4O_9$ instead of bismuth ferrite. As a consequence of an increase of potassium hydroxide concentration, the percentage of $Bi^{3+}$ and $Fe^{3+}$ becomes balanced, and then the BFO phase is formed [50]. When utilizing heavy alkaline NaOH instead of weak KOH for hydrothermal synthesis, both the recipes produce stoichiometric materials. At a pH = 14, a limited quantity of $H_2O_2$ can assist the reduction reaction [42], whereas a milder acidic solution will afford a crude substance with recorded nonstoichiometric $Bi_{12}Fe_{0.63}O_{18.945}$ [42]. Thus, it relies on the techniques used to achieve synthesize the latest phases. Researchers have managed to obtain pure BFO nanoparticles at 200oC by using 7 mol/L KOH and 12 mol/L KOH; however, incorporating NaOH and $LiNO_3$ would either interrupt $Bi_2Fe_4O_9$ and $Bi_{12}(Bi_{0.5}Fe_{0.5})O_{19.5}$phase development or slow it down in the final product no matter the concentration of these reagents [19]. By utilizing potassium nitrate nanoparticles, much smaller nanoparticles of the potassium nitrate content are made. This study suggests that alkali ions ($K^+$, $Na^+$, and $Li^+$) from mineralizer play a significant role in the development of rhombohedral, orthorhombic, and cubic pure bismuth ferrite BFO precipitates, respectively, [51] since the $NO_3$ will raise the supersaturation of the heat transfer solution and intensify the nucleation of the BFO, thus decreasing particle size [52]. Cationic radii for mineralizers also influence the size and morphology of nanoparticles. Hojamberdiev et al. concluded that mineralizers steadily decrease the total particle size before they hit the critical stage. The particle size decreased to exceedingly small amounts [19, 42, 48, 53, 54]; on top of these problems, particle size frequently reaches the limit of 100 nm [19, 53, 54], which compromises the efficiency of the substance. In comparison, the hydrothermal process is expensive and difficult relative to the sol-gel process.

### *2.1.3 Microwave-hydrothermal synthesis*

Microwave-hydrothermal (MH) synthesis is a modified synthesis method based on the hydrolysis technique. The starting materialis very equivalent to the hydrothermal synthesis starting material, but $Na_2CO_3$ is frequently utlized as the mineralizer in conjunction with KOH [2, 3] to gained perovskite $BiFeO_3$ crystals. After transferring the starting material into a Teflon autoclave, it is placed in a microwave oven for the hydrothermal reaction, which is heated at 160-230°C for 30–60 minutes [2, 3, 51, 52]. The particle size produced by MH techniques could be as minimal as 10-50 nm [2] that is smaller than the average particle size produced by hydrothermal process. Furthermore, MH has been utilized to create bismuth ferrite nanotubes with an average size of 50–200 nm and a low bandgap of 2.1 eV [3].

### *2.1.4 Auto-combustion synthesis*

In this process, $Bi(NO_3)_3.5H_2O$ and $Fe(NO_3)_3.9H_2O$ are frequentlyutilized as the oxidizer, whereas glycine ($C_2H_5NO_2$), sucrose ($C_{12}H_{22}O_{11}$), ethylene glycol ($C_2H_6O_2$), ethanolamine ($C_2H_7NO$), citric acid ($C_6H_8O_{7.6}H_2O$), urea ($CON_2H_4$), stearic acid ($C_{18}H_{36}O_2$), etc. may be utlized as fuels. To begin, the metal nitrates are dissolved in dilute $HNO_3$ [55] or distilled water [56] to form an aqueous solution, and then the fuels are added to the solution while constantly stirring. Then, the resulting starting material is moved to a container that can be heated to between 80 and 200 degrees Celsius [55, 56] to form a densify gel-like product. Then, the gels are heated in a furnace to above 300°C to initiate the auto combustion reaction, which produces the final bismuth ferrite nanomaterials [55]. The combustion reactions with glycine and sucrose as fuels are demonstrated below, where bismuth ferrite as the only solid metal oxide product and carbon dioxide, water and dinitrogen simply illuminated during or after the reaction.

$$Bi\,(NO_3)_3 + Fe\,(NO_3)_3 + 3.33C_2H_5NO_2 \rightarrow BiFeO_3 + 6.66CO_2 + 8.33H_2O_{4.66}N_2 \tag{4}$$

$$Bi\,(NO_3)_3 + Fe\,(NO_3)_3 + C_{12}H_{22}O_{11} + 4.5O_2 \rightarrow BiFeO_3 + 12CO_2 + 3\,N_2 + 11H_2O \tag{5}$$

Nevertheless, because secondary phases are frequently constructed with bismuth ferrite, the procedure would be much more complicated. Despite this, Farhadi et al. [57]. Later conducted a study by Yang et al. [55] utilized sucrose ($C_{12}H_{22}O_{11}$) to acquire phase-pure nanoscale BFO powders. The influence of fuel composition on the resulting bismuth ferrite phases discovered that glycine ($C_2H_5NO_2$) or ethanolamine

($C_2H_7NO$) performs better. Throughout overall, the perovskite bismuth ferrite phase is acquired with some secondary phases, making control over the combustion process for phase pure bismuth ferrite difficult.

## 3. Characterization of BFO nanoparticle

To understand and evaluate the potential impacts and harmful effects of nanoparticles (NPS), proper knowledge of their physic-chemical properties is required. This can only be achieved using appropriate techniques to provide all the necessary information about the interested nanoparticles (NPS). Several nanoparticles (NPS) measurement techniques vary in sensitivity and quality information about the sample. The most commonly utilizedmethods including x-ray diffraction (XRD), atomic force microscopy (AFM), transmission electron microscopy (TEM), scanning electron microscopy (SEM), vibrating sample magnetometer (VSM), superconducting quantum interference device (SQUID), Fourier transforms infrared (FT-IR) spectroscopy, energy-dispersive x-ray spectroscopy (EDS), x-ray photoelectron spectroscopy (XPS), Mossbauer spectroscopy (MS), and thermogravimetric analysis (TGA)Several papers have reviewed some of the techniques for the manifestation of ferrite nanoparticles [58–61], as well as the size-dependent functional properties and possible use of BFO nanoparticles are discussed in this section.

### 3.1 X-ray diffraction (XRD)

X-ray diffraction (XRD) is one of the most widely used techniques to differentiate NPs. Typically; XRD provides crystal structure, time-related nature, lattice parameters, and crystal letter size. The following parameter is expected to benefit from Scherrer statistics to extend the most robust XRD size for specific tests. The advantage of XRD techniques, routinely performed on powder-type samples, is usually followed by drying its corresponding colloidal solutions, resulting in common mathematical values with a measure of volume. The consistency of these particles can be determined by comparing the position and strength concerning the peaks and conduct of the guidelines provided by the International Center for Diffraction Data (ICDD, formerly known as the Joint Diversity Guidelines Committee, JCPDS) database. However, it is probably not suitable for non-manufactured materials, and XRD peaks expanded with particles with a size less than 3 nm. Upadhyay et al. found the standard crystallite size of the magnetite NPs using the X-ray line extension and noticed staying at a selection of the number 9–53 nm [62]. The pensiveness of the XRD peaks is caused by particle size/crystallite and lattice types {other than metal extensions [42]. XRD-based size is usually significant compared to so-called magnetic size because of the sub domain names that exist in particles where they always coincide if you look at the same direction and if the particle is a single domain. In contrast, the limited TEM size was more prominent than computer-generated XRD models with larger particles; assuming the particle size is greater than 50 nm, more than one or more crystal boundaries are in their region. XRD cannot distinguish between two parameters; hence, some models' actual (TEM) size can be very noticeable compared to the 50–55 nm determined by the Scherrer formula. Dai and his colleagues set up a very small Au NP that was expected to improve significantly by following ⟨111⟩ (instead compared to ⟨220⟩ one) while the top-to-front analogy was the largest XRD size [43]. Similarly, Li and peers found that when preparing copper telluride nanostructures of different types (i.e., cubes, vessels, and rods), the total stiffness between the associated XRD peaks differed in the number of growing particles [44, 63].

### 3.2 Scanning electron microscopy (SEM)

Scanning electron microscopy (SEM) is a commonly used technique for a high-resolution image of the surface that may separate nanoscale objects. SEM uses electrons in thinking; a bright microscope uses clear light. Mazzaglia et al. blended field-emission SEM (FE-SEM) and XPS standards for analyzing supramolecular colloidal systems in Au NPs/amphiphilic cyclodextrin. Both methods provide essential information about the morphology and nature of the interaction of (Ethylhexyl carbon string) $SC_6NH_2$ and (thiohexadecyl carbon chain) $SC_{16}NH_2$ with Au NPs in the silicon region [19]. Sinclair and colleagues even suggested that SEM and NanoSIMS could help find Au NPs in cells. SEM testing has highlighted its superiority overNanoSIMS when reviewing inanimate NPs in complex biological systems. NanoSIMS delivers low localization of approximately 50 nm, as long as SEM can receive resolutions up to 1 nm. The particles tested have been Raman-active Au-core NPs, and NanoSIMS results in slightly unreadable images in a few cases due to its limited efficiency. Despite this, NanoSIMS introduces new isolation power between isotopes and may or may not be appropriate for Au NPs status [45]. High-resolution SEM (HRSEM) when it comes to rendering Au NPs to cells and cells. The easy appearance of metal NPs is guaranteed by this strategy and by preparing the test quickly and easily. Then again, in biological examples, the requirement to reduce the recycling of materials may form the appropriate metal layer; for this reason, it increases the risk of radiation damage when it comes to specimens. The advantage of HRSEM, when compared to other photographic methods, is the ability to lower precision and analyze the views of nanometric elements with their broader perspective. It makes it possible to review the specific spatial plan of the NPS and assess the functional relationship between them. The study results praised HRSEM's potential as a moderately advanced tool to effectively highlight points that enhance Au NPs transmission through the epidermis buffer. It can be observed on a powerful and versatile device to capture better the interactions between biological systems and metallic nanostructures [46]. SEM size and AFM are compared to the same set of NPs, i.e., $SiO_2$ and Au NPs in mica or silicon substrates. For example, AFM information has enabled the magnitude of the nano object to sub-nanometric accuracy, yet the dimensions of the sides (corresponding to the X and Y axes) have significant errors due to tip/sample flexibility. Compared to AFM, SEM cannot provide any metrological information about the length of this NPS; but even then, modern SEM can provide the correct measurements for their back weight. The size of the circular $SiO_2$ NPs using both techniques yielded very similar results, showing consistent consistency and resistance of both metals [64]. SEM can be managed by looking at the transmission mode, that is, by using a method that is perceived as transfer to an electron scanning microscope (T-SEM) (see **Figure 1**). The most advanced NP tests can be taken by finding the full facts and studying NP practices if you look at the transmission mode. By Rades et al., A combination of different techniques such as SEM, T-SEM, EDX, and Auger microscopy (SAM) scanning is an excellent way to assess the depth of morphological and chemical properties specific. Silica and titania NPS. While acknowledging that, methods such as SAXS, DLS, XPS, XRD, and BET may be appropriate to define NPs ensembles, except for particles alone. T-SEM enables rapid analysis of NP form, although its lateral quality is limited to NP magnitude up to 5–10 nm. TEM offers excellent high-quality images, but T-SEM can be easily integrated with EDX to assess NP size and much-needed structure [47] quickly.

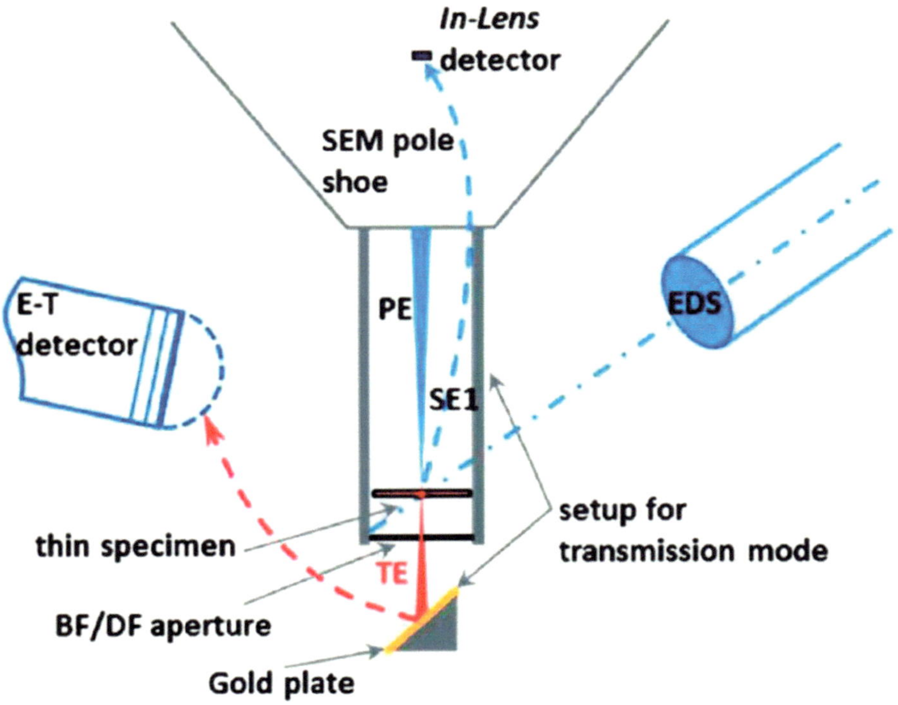


**Figure 1.**
*Picture of an SEM/EDS technique running when you look at the transmission setting using the Zeiss single-unit transmission install(PE: Primary electrons; SE1: Secondary electrons emitted during the point of the effect regarding the PEregarding the sample; TE: Transmitted electrons; BF: Bright field; DF: Dark field; E-T; Everhart-Thornley detector). Reprinted with authorization from Ref. [47]. Copyright Royal Society of Chemistry 2014.*

### 3.3 Transmission electron microscopy (TEM)

Transmission Electron Microscopy is a straightforward notification process for monitoring atomic and molecular alignment. TEM is an effective investigation into the size and shape of nanoparticles. Jewelry and particle measurements can be found in the TEM test. TEM is a microscopy method that uses a link between a continuous electron beam current (i.e., power is usually within the range of 60 to 159 keV) as well as a small test. As soon as the electron beam reaches the test, the electron element is sent, although sleep is transmitted flexibly or in elastically. The size of the connection depends on various problems, such as size, test density, and initial structure. The final image was made using information obtained by transmitting electrons. TEM is the most common method for measuring nanoparticle size and composition as it provides precise images for testing and more accurate testing of nanoparticle homogeneity. However, a few limitations should be considered fixed whenever production of this method is produced, such as the problem of measuring a few particles or unreliable images due to the impact of the stand. Whenever examples are surprisingly similar, alternatives that determine the sheer number of nanoparticles can provide more reliable results, such as the SAXS of the most potent formula and the Scherrer [48], or the XRD presentations and the Scherrer formula. Nanoparticle structures almost certainly depend not only on their size and shape but also on something else, such as particle lengths. For example, whenever two metal nanoparticles are helped to bring it closer, their few plasmons change their plasmonic arrangement and change their color. Later, TEM features have been is used to describe the nanoparticle collection of various biomedical systems, including hearing and diagnosis, when based on the presence of a biomarker or analyst [49, 50]; treatment, in which the collection results in the therapeutic development of the nanoparticle effects [51]; and (iii) photographs, in which the collection shows the development of a response mark [52]. In order to obtain reliable results, additional procedures should be taken to suspend testing if it is considered that an inadequate code of conduct may result in test modification or artifact production [55]. The primary control of nanoparticle-setting settings has been achieved for many years now, and it has been successful} NP installs can lead to NP dinner directors. Subsequent construction of various nanocrystals results in new multidisciplinary structures that include features of specific emerging barriers and an increase in recent and attractive structures [56]. TEM-elements are among the mechanisms used to introduce assorted superlattice nanocomposites, which are usually isostructural in a wide range of atomic crystal systems [57]. These brand new triangular compounds are created by various NPs (e.g., quantum dots, metals, and magnetic NPS), and their final field} and structure can be handled by creating a colloid circuit [57] or a direct link to DNA [55].

### 3.4 Superconducting quantum interference device magnetometry

Superconducting quantum interference device magnetometry (SQUID) is a tool for calculating the magnetic fields of nanoscale objects. Nanomaterials, in some cases, exhibit properties that are different from those in the plural form due to their small size and awareness of local conditions. As the material decreases in size, it progresses from multiple domains to a single domain and eventually to a superparamagnetic state. Standard SQUID measurements produce features such as magnetization filtration (MS), magnetization remanence (Mr.), and inhibitory temperature (Tb). On the side of NPs, the magnetic response of particular molecules can be tested by SQUID. A magnetic scanner that incorporates nanoSQUID features that have already been created recently made about a sharp quartz apex. NanoSQUID is considered the most inspiring investigation into nanoscale magnetic imaging and spectroscopy. The nanoSQUID sensor requires a Josephson subcutaneous micron coating, provided by two Dayem nano bridges (nano-constriction of the superconducting film), made of electron beam lithography {focused ion beam (FIB) with the same length and width and compliance length. The main SQUID requirement developed to obtain a magnetic NPS is a small SQUID area. Loop sizes should prefer the NPS integrated [58] to achieve the most fantastic merging feature. About magnetic resonance | microscopy or magneto-optic spin recognition, nanoSQUID provides the advantage of the exact magnitude of the magnetic field in small spin paths. In addition to its inefficient construction with the step of nanopatterning alone, Dayem nanoSQUID bridges can withstand the magnetic field used inside the SQUID loop plane [59]. The test set of the nanoSQUID is shown in **Figure 2**.

### 3.5 Vibrating sample magnetometry (VSM)

Vibrating test magnetometry (VSM) has become an additional feature that can be used to report that MH will hinder magnetic nanomaterials and get parameters like Ms. the magnetic field, temperature, and duration. Fabris and his colleagues made NP magnets controlled by the rainfall system in the presence of tetramethylammonium hydroxide. All tested samples were superparamagnetic, evidenced by

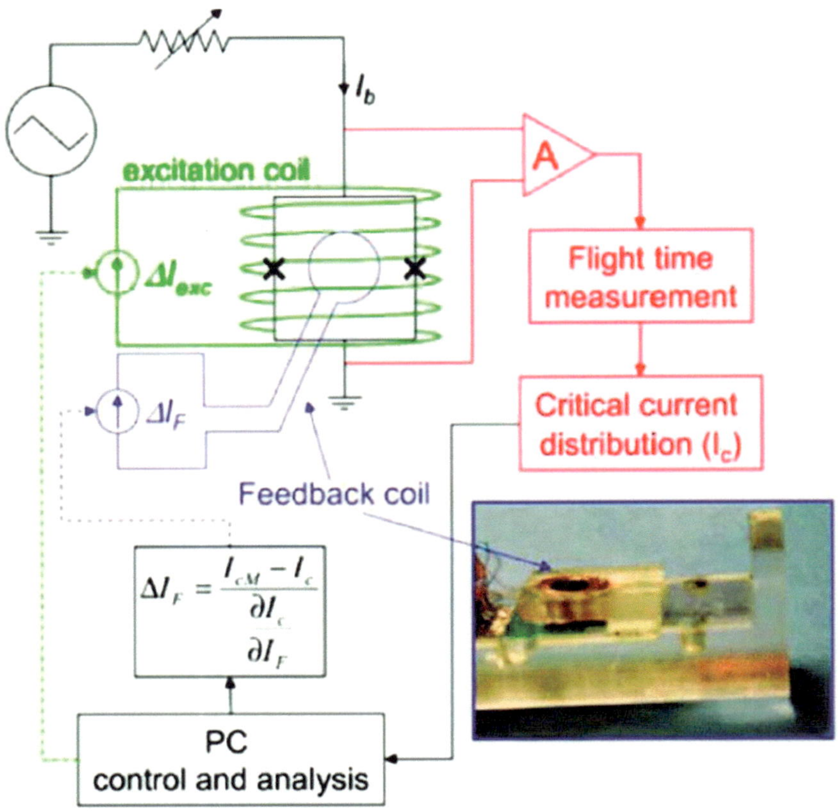


**Figure 2.**
*Scheme of the experimental setup for the NP magnetization dimensions. The difference of the crucial current is acquired by averaging the changing current activities assessed simply by using a time-of-flight method. The quality of the essentialcurrentmeasurementsis mostly about 1 component in 104. The feedback circuit permits the rise of the robust linear variety of the sensor. The image shows the owner, the sample, andthe multiturn feedback coil. Reprinted with authorization from Ref. [59]. Copyright 2013 springer.*

zero compulsions and zero residues regarding the magnetization loop. Saturation magnetization retains the purpose of the NP size line [60]. Kumari et al. define the concept of reversible curves in the first order, which is the setting of hysteresis barriers from the main loop. Curve reversal order (FORC) works well in determining domain size, composition, and interaction on a magnetic field. It is an effective technique to obtain a semi-quantitative measure of the sufficient size of a magnetic particle. Under certain circumstances, the FOC may facilitate the disclosure of the existence of secondary magnetic experiments, thus assisting in the formation of more accurate characters for magnetic structures. VSM magnetometry was eventually used} to obtain such FOC measurements [61]. For further work, FeCo NPs used a framework of long anisotropic chains that had been prepared with a sputter-based gas cooling process, and VSM tested its magnetic properties. Deep interactions involving exchanges between NPs were confirmed in a chain-like sample, while well-dispersed samples revealed a completely different magnetic effect. ZFC/FC curves and time-dependent magnetic field measurements were assisted to obtain data regarding the thermal stability of the tested NPs [65]. In addition to the entire structures of the hysteresis loop of magnetic media, there has been a growing desire for the balance of remanence curves. Remanence measurement determines only the irreversible part of magnetization and, for that reason, makes it possible for the transformation events to be removed from the hysteresis scale, which includes the retractable object. Two primary reversal curves create isothermal remanence and a DC demagnetization curve. The previous one is being evaluated following the use and completion of a field with all magnetic resonance samples. The DC demagnetization curve is calculated based on the state of focus using the increasing fields of magnetic removal. VSM scales can detect these remanence curves and provide an accurate distribution of the neglected field of building materials [66].

### 3.6 X-ray photoelectron spectroscopy (XPS)

X-ray photoelectron spectroscopy (XPS) is one of the most widely investigated research methods for further chemical reactions, it is also used to classify nanoscale objects. What is clear from physical thought is the effect of electrical images [67]. XPS is an effective quantitative process for determining elements' electronic structure, primary structure, and oxidation conditions in a matter. It can also test the ligand exchange interactions with local NP operations and key structures/shells and operate under very high vacuum conditions. Nag and colleagues have published a review paper outlining the role of XPS as a fun way to study the internal heterostructures of NPs. For example, it was used to investigate the structure of a crystal structure dependent on the environment of the metal chalcogen NPs of various sizes [67]. It can also distinguish between core/shell and homogeneous alloy structures and identify ligand binding modes such as trioctylphosphine oxide (TOPO) on the surface of metal chalcogenide NPs. For example, if a TOPO bond is preferable to the surface of the metal, then a portion of the excess chalcogenide may be easily oxygenated in the air. In contrast to microscopy techniques, such as TEM, which uses side-by-side alignment to determine elements in a straight line to the test electron column, XPS investigates the formation of an object aligned with an electron line. Concerning the core-shell NPs, shard has published a text that reports the precise way to translate XPS data for those types of particles. It involves a straightforward approach to turning XPS firmness into an overlay layer, ideal for round NPs. As an additional benefit of XPS, the author mentions that it provides in-depth information, such as the size of NPs (at a depth of 10 nm from the top), and does not seriously damage the samples. Two barriers to XPS analysis are sample preparation (i.e., solid dry form is required without contamination) in addition to the information definition.

### 3.7 Fourier transform infrared spectroscopy (FTIR)

Fourier transform infrared spectroscopy (FTIR) measures the absorption of electromagnetic radiation in the infrared medium (4000–400 $cm^{-1}$). When a molecule absorbs infrared radiation, the moment of the dipole is somehow altered, and the molecule becomes an active IR. The recorded spectrum contains energy-related bands, bonds, and specific functional groups that provide molecular structures and interactions. Feliu and colleagues used the combined in situ ATR-FTIR and defaulting electrochemical mass spectroscopy to investigate the effect of Pt nanostructures during ethanol oxidation (DEMS). These methods assisted in the electrochemical analysis of adsorbates and the detection of flexible reaction products. Their findings support previous findings, showing that the decomposition products selected were related to the above structures, the formation of Coads in the (100) domains, and the formation of acetaldehyde/acetic acid in 111 domains. One study used carbon-backed NPs (3–8 nm in size) to obtain CO oxidation and a catalytic process was performed with DRIFTS and quadrupole mass spectrometry (QMS). According to the QMS results, FTIR ratings of adsorbed CO verified Coad and Oad variations in different test phases. While changes in the distribution of CO over different types of Pt sites were also observed. Overall, DRIFTS was considered an essential tool for assessing the local structure of Pt NPs in situ. Shukla et al. published a paper on FTIR oleylamine research. The anterior ligand is bound to FePt NPs in both monodentate and bidentate forms, while oleylamine is bound to FePt and the NH2 group is substantial. In addition, Au/Ag bimetallic NPs are dodecanetiol-soluble and dissolved insoluble solvents incorporated into water/toluene via a two-phase synthetic route [43]. The most important finding from the XPS and FTIR ratings was that Ag atoms were enriched on the outer edge of the hybrid clusters compared to Au atoms. ATR-FTIR was used in another study to assess the effect of Ag NP content on photocatalytic degradation of oxalic acid advertised in TiO2 NPs. Various Ag NP values were tested, and it was found that inserting only a tiny amount (2 percent) significantly improved the photocatalytic performance of TiO2 NPs. NP composite films' location and chemical structure/composition were observed using AFM and XPS. Tzitzios et al. created hexagonal Ni NPs with a diameter of 13–25 nm by reducing nickel stearate in the presence of PEG, oleic acid, and oleylamine. The presence of different groups on the surface of NPs, such as HCvCH- setting in OAC and OAm, was indicted by the FTIR spectra, and the mechanisms for ligand binding in the NP area were also investigated. Haram and its allies have used the hot-shot method to make copper zinc tin sulpho-selenide ($CZTS_xSe_{1-x}$) nanocrystals. For incorporation, the precursors were dispersed in OAm and heated to T > 200°C. OAm advertising on particle surfaces is reflected in FTIR ratings. Feature belts are derived from components present in the OAm molecule and effectively interact with NPs. The number of NP organic ligands is determined [67].

## 4. Properties in BFO nanoparticles

### 4.1 Ferroelectric

Ferroelectricity electricity is usually expected to disappear in smaller sizes due to a decrease in the relative length of the dipoles [66, 67]. Exact measurement of ferroelectricity in 0-D nanostructures and identification of proper ferroelectric size - the result is a difficult task. To begin with, making electronic contact with a single nanoparticle can only be done using scanning test methods, and no report has ever published the ferroelectric hysteresis characteristics of a single nanoparticle according to our knowledge. The problem is exacerbated by the small size and leaky nature of BFO nanoparticles due to the reduced bandgap and switching voltages that are likely to be very close to the dielectric cracking [46]. Vasudevan et al. [57] investigated the ferroelectric characterization of BFO nanoparticle clusters prepared for automatic firing methods using band excitation piezoresponse spectroscopy (BEPS) and piezoresponse force microscopy (PFM) (larger than 50 nm). They confirmed the ferroelectricity of nanoparticles by obtaining a symmetric piezoresponse loop with a compliant 8 V voltage of a single batch distributed in the LSMO/STO substrate (**Figure 3 (a)**). In addition, they found the properties of the ferroelectric domain (**Figure 3 (b)** and **(c)** within groups of particles, similar to those found in small BFO films [61]. There is often a direct link between ferroelectricity and lattice strain (using a known strain-polarization coupling). This means that a detailed structural investigation into each area's lattice parameter or removal can

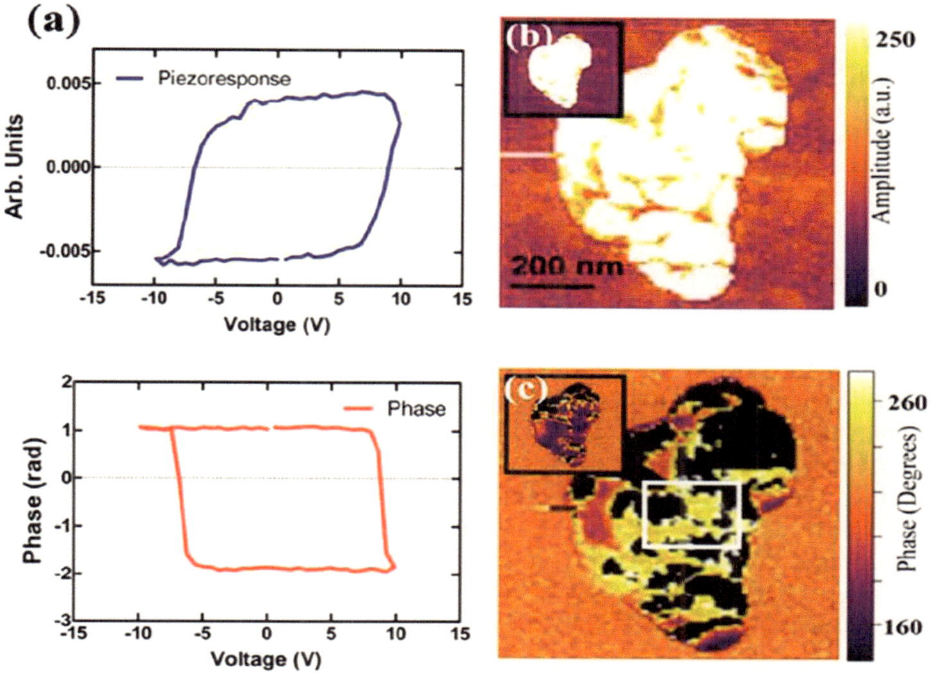


**Figure 3.**
*(a) Piezoresponseand phase hysteresis loops of a solitary BiFeO$_3$ nanoparticle. Out-of-plane PFM amplitude (b) and phase (c) pictures of a nanoparticle cluster, pre, and post-putting on +10 V, 5 s pulse towards the center of this cluster. Insets in (b) and (c) demonstrate the PFM amplitude and phase pre applying the bias; correspondingly, the above assessment had been performed.*

provide important information. Selbach et al. [31] A systematic study of the relationship between nanoparticle size and BFO lattice parameter illustrates this point. As shown in **Figure 3**, these researchers found that nanoparticles more prominent than 30 nm have lattice limits than a BFO mass. In contrast, at less than 30 nm, the lattice parameter of BFO nanoparticles extends from the mass and approaches the cubic (i.e., paraelectric) perovskite structure. There is a decrease in the rhombohedral deviation of the cell unit (i.e., a decrease in c/a). When the rhombohedral angle reaches 60o, it equals the unit, indicating a fine cubic perovskite [31]. A significant dc size of 9 1 nm of ferroelectricity was obtained using the empirical model to match tetragonality based on BFO size [31]. Automatic polarization determined by removing $Bi^{3+}$ and $Fe^{3+}$ cations at 13 nm nanoparticles was 75% of the total volume. This makes BFO nanoparticles an exciting class for many materials because they can have both a strong magnetic field (discussed below) and sufficient ferroelectric polarization for novel applications [31].

### 4.2 Photocatalytic

Compared to BFO thin films, which typically have a bandgap of 2.7 eV [22], BFO nanoparticles prepared chemically have a bandgap as low as 1.8–2.3 eV. As a result, they are appealing for use in photocatalysis. Nanosized BFO particles have demonstrated improved photocatalytic performance, which can be applied to the degradation of organic pollutants such as dye compounds of Methyl orange (MO), Methylene Blue (MB), Congo Red (CR), or Rhodamine B. (RhB). Geo et al. [22], for example, confirmed that BFO nanoparticles, in addition to responding to UV light, have excellent MO degradation ability when exposed to visible light (**Figure 4(a)**).As shown in **Figure 4(b)**, Geo et al. discovered that Gd-doped BFO nanoparticles could improve their photocatalytic properties by increasing RhB degradation rates from 79 percent for BFO to 94 percent $Bi_{0.9}Gd_{0.1}FeO_3$ [26]. Using BFO as a photocatalytic agent is its photostability, affecting photocatalytic efficiency under visible light. It is investigated the nonphotostability of BFO nanoparticles by studying RhB dye decolorization at pH 2, 4, and 6, 7, as shown in **Figure 4(c)**. They discovered that photo corrosion occurs in the RhB dye solution due to the dissolution of Fe from the Fe-O bond, resulting in nonphoto stability. This photo corrosion can be explained as an offshoot of the BFO band offset concerning the RhB dye, in which holes can be injected from the RhB dye into the BFO valence band. As shown in **Figure 4**, replacing the BFO nanoparticles in the RhB solution at regular intervals can achieve a much higher decolorization rate, which can even exceed TiO2 (d). As a result, BFO nanoparticles are extremely promising for visible-light-driven photochemistry. Compared to small BFO films, typically with a 2.7 eV bandgap [22], BFO nanoparticles are chemically modified with a band as low as 1.8–2.3 eV [22, 54]. As a result, they requested use in photocatalysis. Nanosized BFO particles have shown improved photocatalytic activity, which can be used in the decomposition of organic pollutants such as compounds of Methyl orange (MO), Methylene Blue (MB), Congo Red (CR), or Rhodamine B (RhB).

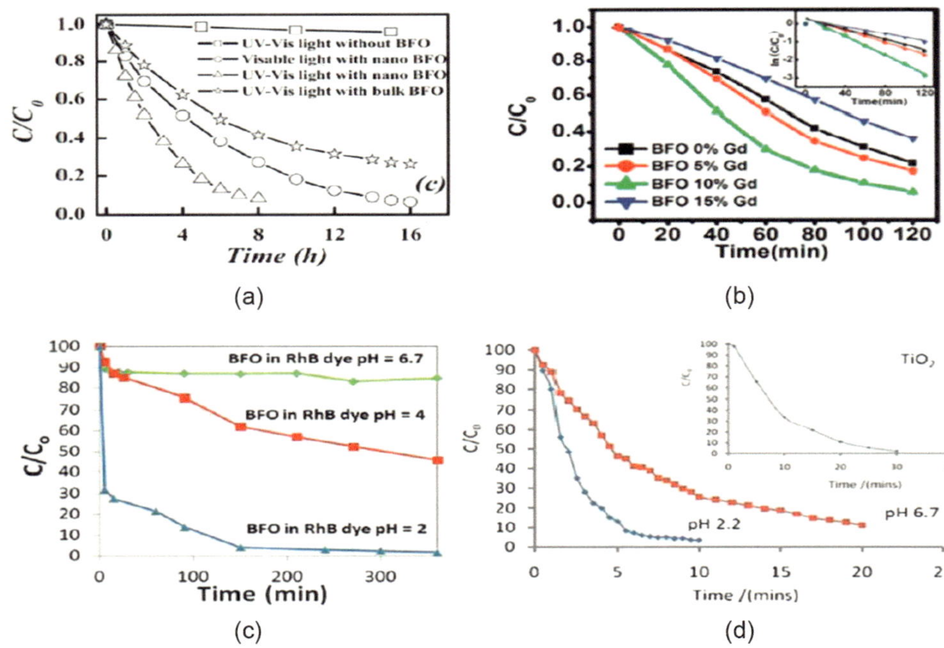


**Figure 4.**
*(a) Photocatalytic removal activities of methyl orange under UV-vis light irradiation and visible light irradiationutilizingBiFeO$_3$ nanoparticles as well as bulk, [51] recreated with authorization from ref 104, copyright2007 WILEY-VCH Verlag GmbH & Co. kraal, Weinheim; (b) photocatalytic removal effectiveness of RhButilizingGd substituted BFO nanoparticle samples, [26] recreated with authorization from ref 26, copyright 2010 American Chemical Society (c) and (d) photo removal information of RhB utilizingBiFeO$_3$ as photo reagent at different pH values under AM1.5 lighting; [65] (d)tests had been carried outwith the substitution of theBFO nanopowders at frequentperiods, into the dye answer, decolorization at pH = 2 reveals higher than 95% decolorization after 10 min, the inset demonstratesthe removal activities of RhB utilizingnanostructured (Degussa P25) TiO2, [65] recreated with authorization from ref 103, copyright 2012 Royal Society of Chemistry.*

Geo et al. [22], for example, confirmed that BFO nanoparticles, in addition to UV radiation, have excellent potential for MO degradation when exposed to visible light (**Figure 4 (a)**). As shown in **Figure 4 (b)**, Geo et al. found that Gd-doped BFO nanoparticles can enhance their photocatalytic properties by increasing the rate of RhB degradation from 79 percent in BFO to 94 percent in $Bi_{0.9}Gd_{0.1}FeO_3$ [26]. As a result, there is strong pressure to develop alternatives. Due to the p-n junction built into the p-type BFO interface and the n-$TiO_2$ type, the presence of BFO nanoparticles improves the image degradation performance under visible light while preventing the reunification of electron-generated electrons. As shown in **Figure 4 (a)**, the efficiency of the depletion of $TiO_2$ nanofibers in MB is only 3% after 150 minutes, but this performance can be achieved to nearly 100% when 5 mol% and 10 mol% $BiFeO_3$ nanoparticles are incorporated in $TiO_2$ nanofibers [64]. Moreover, boosting the density of BFO nanoparticles has a negative impact on photocatalytic activity because they can obscure active sites of $TiO_2$ and inhibit electron transfer to the BFO / TiO optical connector. Correspondingly, under visible light, the performance of BFO / $TiO_2$ nanotubes image conversion can be increased from 0.7 percent for pure $TiO_2$ nanotubes to 3.2 percent for BFO/$TiO_2$ composite nanotubes [50], which can be seen in **Figure 4 (b)**. Other abnormalities which have been encountered to improve photocatalytic function include BFO-graphene and nanohybrid [52]. The improved magnetic performance and optical illumination of the BFO nanoparticles described above make them excellent candidates for advanced development and application.

### 4.3 Magnetoelectric coupling

Magnetoelectric bonding (ME), which can detect coexistence with the same combination of ferroelectric and magnetic properties, maybe the most notable feature of nano-BFO-based materials. As a result, many researchers are focusing on the promise of merging the boundaries of magnetic and electrical order. Zhao et al. [3] were the first to demonstrate the electrical control of the antiferromagnetic domain structure in a single-phase BFO film, showing a solid interaction between the two types of structure at room temperature. They found that before and after electric cooling, the antiferromagnetic base structure was firmly attached to the ferroelectric base structure. Automatic polarization occurs near the axis. The angles 71°, 109°, and 180° show three distinct divisions as the rhombohedral axis changes. It was found that the interaction between the antiferromagnetic domain and the ferroelectric domains occurs only with a polarization of 71° and 109°, but not with a change of 180o ferroelectric polarization. This work is an essential first step for researchers interested in investigating ME nano-BFO compound materials. However, due to the slight spin canting, it is not easy to achieve the significant coefficient of magnetoelectric coupling magnetization. The acquisition of magnetic anisotropy in the optical connector of ferromagnetic-antiferromagnetic heterostructures allows for a more extraordinary animation of device applications. Thin-film heterostructures also benefit from the way magnetic anisotropy of the system can be constructed with epitaxy. As a result, researchers are keenly interested in combining ferromagnet-multiferroic exchange heterostructures for BFO sub-film, particularly oxide ferromagnet/BFO heterostructures and transition metal ferromagnet/BFO heterostructures. For oxide ferromagnet/BFO heterostructures, $La_{0.7}Sr_{0.3}MnO_3$ (LSMO) is popular. It is found that different magnetic exchange effects were produced at different LSMO/BFO sites because $Fe^{3+}$ and $Mn^{3+}$ or $Mn^{4+}$ are ferromagnetic, competing with a wide range of antiferromagnetic order. Later, Wu et al. [19] used the LSMO/BFO system to demonstrate the electronic control of exchange bias. They found a reversible shift between the two biased regions by changing the ferroelectric polarization of the BFO. This is a significant step forward in controlling pornography by power control, and it is an essential step towards spintronic electronic control devices. However, due to the relative temperature dependence of the ferromagnet/BFO heterostructure exchange, achieving renewable electrical control of the magnetoelectric junction at room temperature remains difficult. The ferromagnet transition metal ($Co_{0.9}Fe_{0.1}$) ferromagnetic layer to form a heterojunction with the BFO to control the local magnetic field room temperature. Magnetic anisotropy was modified using an electric field in the aircraft. After an electrical switch was investigated, they found that the linear domains moved from left to right. The domains then recede after the use of a different electric field. At room temperature, the electric field can control the flexible wall of the electric field, as shown in this series of pictures. Most importantly, this work demonstrated the magnetoelectric in front of the electric field at room temperature.

## 5. Conclusions

In conclusion, the nano-BFO is a suitable multiferroic nanomaterial with a few unique features, including high polarization of residue, magnetoelectric composite at room temperature, and small bandgap. It also serves as a flexible platform for exploring new novel works. This review of the topic focuses on significant improvements made in the study of integrating nano-BFO-based materials, features, structures, and applications. Several operational strategies have been proposed. Proper substrate selection and morphotropic phase boundary formation have been widely used in the modification to improve ferroelectric efficiency and magnetic field. The impact of size on behavior on nano-BFOs is also a significant problem.

Efforts to achieve outstanding results have been made on fixed memory, piezoelectric sensors, and photodetectors. However, many challenges remain until those modern facilities are completed. First and foremost, it is essential to improve the efficiency of ferroelectric and magnetoelectric energy. The nano-BFO is highly fragmented. Is it possible to achieve high polarization of ferroelectric power? Will nano-BFO devices show a broad magnetoelectric response at room temperature? Second, should sufficient switching speed be required for systems based on ferroelectric or ferromagnetic switching? Most of the recorded speed of change and the durability of the switch does not meet the memories of the universe. As a result, there has been a long way to go to increase memory on nano-BFO-based devices. Third, the physiological processes under the influence of photovoltaic and photocatalytic activity in nano-BFO remain a mystery. BFO photocatalyst and photovoltaic devices are expanded only by exposing the visible structures. Several ideas have been suggested; however, there is much disagreement, and further research is needed. Fourth, only a few nano-material structures have been observed to reflect composite structures. Many nanomaterial systems should be built to investigate the potential for new applications. With this review, we hope to provide a current summary of the problems and opportunities that arise, which may inspire more researchers to pursue future nano-BFO production.

*DOI: http://dx.doi.org/10.5772/intechopen.104777*

## Author details

Sheela Devi[1*], Venus Dillu[2] and Mekonnen Tefera Kebede[3]

1 Department of Applied Sciences, Maharaja Surajmal Institute of Technology, New Delhi, India

2 Departement of Physics, Sharda University, School of Basic Science and Research, Greater Noida, UP, India

3 Departement of Physics (Nanotechnology), Sharda University, School of Basic Science and Research, Greater Noida, UP, India

*Address all correspondence to: physics.sheela@gmail.com

## References


[1] Zhu X, Liu Z, Li Z, Yu T. Microwave hydrothermal synthesis, structural characterization, and visible-light photocatalytic activities of single-crystalline bismuth ferric nanocrystals. Journal of the American Ceramic Society. 2011;**94**:2688-2693. DOI: 10.1111/j.1551-2916.2011.04430.x

[2] Joshi UA, Jang JS, Borse PH, Lee JS. Microwave synthesis of single-crystalline perovskite $BiFeO_3$ nanocubes for photoelectrode and photocatalytic applications. Applied Physics Letters. 2008;**92**(24):1-4

[3] Zhao T, Scholl A, Zavaliche F, Lee K, Barry M, Doran A, et al. Electrical control of antiferromagnetic domains in multiferroic $BiFeO_3$ films at room temperature. Nature Materials. 2006;**5**(10):823-829

[4] Park T, Papaefthymiou GC, Viescas AJ, Moodenbaugh AR, Wong SS. Size-dependent magnetic properties of $BiFeO_3$ nanoparticles. Nano Letters. 2007;**7**(3):766-772 [Internet] Available from: http://pubs.acs.org/doi/abs/10.1021/nl063039w

[5] Zhang H, Kajiyoshi K. Hydrothermal synthesis and size-dependent properties of multiferroic bismuth ferrite crystallites. Journal of the American Ceramic Society. 2010;**93**(11):3842-3849

[6] Das RR, Kim DM, Baek SH, Eom CB, Zavaliche F, Yang SY, et al. Synthesis and ferroelectric properties of epitaxial $BiFeO_3$ thin films grown by sputtering. Applied Physics Letters. 2006;**88**(24):4-6

[7] Mazumder R, Sujatha Devi P, Bhattacharya D, Choudhury P, Sen A, Raja M. Ferromagnetism in nanoscale $BiFeO_3$. Applied Physics Letters. 2007;**91**(6):1-3

[8] Nath R, Hong S, Klug JA, Imre A, Bedzyk MJ, Katiyar RS, et al. Effects of cantilever buckling on vector piezoresponse force microscopy imaging of ferroelectric domains in $BiFeO_3$ nanostructures. Applied Physics Letters. 2010;**96**(16):1-4

[9] Park M, Hong S, Klug JA, Bedzyk MJ, Auciello O, No K, et al. Three-dimensional ferroelectric domain imaging of epitaxial $BiFeO_3$ thin films using angle-resolved piezoresponse force microscopy. Applied Physics Letters. 2010;**97**(11):3-5

[10] Au M, Rose J, Bottero J, Lowry G V, Jolivet J, Wiesner MR. Towards a definition of inorganic nanoparticles from an environmental, health and safety perspective. Nature Nanotechnology. 2009;**4**:634-641. DOI: 10.1038/nnano.2009.242

[11] Manikandan A, Sridhar R, Antony SA, Ramakrishna S. A simple aloe vera plant-extracted microwave and conventional combustion synthesis: Morphological, optical, magnetic and catalytic properties of $CoFe_2O_4$ nanostructures. Journal of Molecular Structure. 2014;**1076**:188-200. [Internet] Available from:. DOI: 10.1016/j.molstruc.2014.07.054

[12] Eatemadi A, Daraee H, Karimkhanloo H, Kouhi M, Zarghami N, Akbarzadeh A, et al. Carbon nanotubes: Properties, synthesis, purification, and medical applications. Nanoscale Research Letters. 2014;**9**(1):393 [Internet] Available from: http://www.ncbi.nlm.nih.gov/pubmed/25170330%0Ahttp://www.pubmedcentral.nih.gov/articlerender.fcgi?artid=PMC4141964

[13] Wang L, Xu J, Gao B, Chang A, Chen J, Bian L, et al. Synthesis of BiFeO

3 nanoparticles by a low-heating temperature solid-state precursor method. Materials Research Bulletin. 2013;**48**(2):383-388. [Internet] Available from:. DOI: 10.1016/j.materresbull.2012.10.038

[14] Ghosh S, Dasgupta S, Sen A, Maiti HS. Low temperature synthesis of bismuth ferrite nanoparticles by a ferrioxalate precursor method. Materials Research Bulletin. 2005;**40**(12):2073-2079

[15] Chen C, Cheng J, Yu S, Che L, Meng Z. Hydrothermal synthesis of perovskite bismuth ferrite crystallites. Journal of Crystal Growth. 2006;**291**:135-139. DOI: 10.1016/j.jcrysgro.2006.02.048

[16] Tang NJ, Zhong W, Jiang HY, Wu XL, Liu W, Du YW. Nanostructured magnetite ($Fe_3O_4$) thin films prepared by sol-gel method. Journal of Magnetism and Magnetic Materials. 2004;**282**:92-95. DOI: 10.1016/j.jmmm.2004.04.022

[17] Muneeswaran M, Jegatheesan P, Giridharan NV. Synthesis of nanosized $BiFeO_3$powders by co-precipitation method. Journal of Experimental Nanoscience. 2013;**8**(3):341-346

[18] Safarik I, Horska K, Pospiskova K, Maderova Z, Safarikova M. Microwave-assisted synthesis of magnetically responsive composite materials. IEEE Trans. Magn. 2013;**49**(1):213-218. DOI: 10.1109/TMAG.2012.2221686

[19] Wu W, Wu Z, Yu T, Jiang C, Kim W. Recent progress on magnetic iron oxide nanoparticles: synthesis, surface functional strategies and biomedical applications. Science and Technology of Advanced Materials. 2015;**16**:023501. DOI: 10.1088/1468-6996/16/2/023501.

[20] Kim JK, Kim SS, Kim WJ. Sol-gel synthesis and properties of multiferroic $BiFeO_3$. Materials Letters. 2005;**59**:4006-4009. DOI: 10.1016/j.matlet.2005.07.050

[21] Gao F, Yuan Y, Wang KF, Chen XY, Chen F, Liu JM, et al. Preparation and photoabsorption characterization $BiFeO_3$ nanowires. Applied Physics Letters. 2006;**89**(10):3-5

[22] Gao BF, Chen X, Yin K, Dong S, Ren Z, Yuan F, et al. Visible-light photocatalytic properties of weak magnetic $BiFeO_3$ nanoparticles. Advanced Materials. 2007;**19**:2889-2892. DOI: 10.1002/adma.200602377

[23] Chakrabarti K, Das K, Sarkar B, Ghosh S, De SK, Sinha G, et al. Enhanced magnetic and dielectric properties of Eu and Co co-doped $BiFeO_3$ nanoparticles. Applied Physics Letters. 2012;**101**(4):1-6

[24] Qian FZ, Jiang JS, Guo SZ, Jiang DM, Zhang WG. Multiferroic properties of $Bi_{1-x}Dy_xFeO_3$ nanoparticles. Journal of Applied Physics. 2009;**106**(8). DOI: 10.1063/1.3245390

[25] Kang Y, Cao M, Yuan J, Shi X. Microwave absorption properties of multiferroic $BiFeO_3$nanoparticles. Materials Letters. 2010;**63**(15):1344-1346. [Internet] Available from:. DOI: 10.1016/j.matlet.2009.03.010

[26] Guo R, Fang L, Dong W, Zheng F, Shen M. Enhanced photocatalytic activity and ferromagnetism in Gd doped $BiFeO_3$ nanoparticles. Journal of Physical Chemistry C. 2010;**114**(49):21390-21396

[27] Liu J, Fang L, Zheng F, Ju S, Shen M. Enhancement of magnetization in Eu doped $BiFeO_3$ nanoparticles. Applied Physics Letters. 2009;**95**(2):2-5

[28] Lotey GS, Verma NK. Structural, magnetic, and electrical properties of Gd-doped $BiFeO_3$ nanoparticles with reduced particle size. Journal of Nanoparticle Research. 2012;**14**(3):742 [Internet] Available from: http://link.springer.com/10.1007/s11051-012-0742-7

[29] Arora M, Sati PC, Chauhan S, Chhoker S, Panwar AK, Kumar M. Structural, optical and multiferroic properties of $BiFeO_3$ nanoparticles synthesized by soft chemical route. Journal of Superconductivity and Novel Magnetism. 2013;**26**(2):443-448

[30] Chakrabarti K, Das K, Sarkar B, De SK. Magnetic and dielectric properties of Eu-doped $BiFeO_3$ nanoparticles by acetic acid-assisted sol-gel method. Journal of Applied Physics. 2011;**110**(10). DOI: 10.1063/1.3662178

[31] Selbach SM, Tybell T, Einarsrud M, Grande T. Size-dependent properties of Multiferroic $BiFeO_3$ nanoparticles. Chemistry of Materials. 2007;(19):6478-6484. DOI: 10.1021/cm071827w

[32] Courses C, Ianculescua A, Gheorghiub FP, Postolacheb P, Opreaa O, Jaiswal A, et al. References 1. 2011;**2**: 179-218.

[33] Huang F, Wang Z, Lu X, Zhang J, Min K, Lin W, et al. Peculiar magnetism of $BiFeO_3$ nanoparticles with size approaching the period of the spiral spin structure. Scientific Reports. 2013;**3**:1-7

[34] Amin M, Rafique HM, Yousaf M, Ramay SM, Saleem M, Abbas SK, et al. Multiferroicity in sol-gel synthesized Sr/Mn co-doped $BiFeO_3$ nanoparticles. Journal of Materials Science: Materials in Electronics. 2017;**28**(22):17234-17244

[35] Wang X, Lin Y, Ding X, Jiang J. Enhanced visible-light-response photocatalytic activity of bismuth ferrite nanoparticles. Journal of Alloys and Compounds. 2011;**509**(23):6585-6588. [Internet] Available from:. DOI: 10.1016/j.jallcom.2011.03.074

[36] Wang X, Lin Y, Zhang ZC, Bian JY. Photocatalytic activities of multiferroic bismuth ferrite nanoparticles prepared by glycol-based sol-gel process. Journal of Sol-Gel Science and Technology. 2011;**60**:1-5. DOI: 10.1007/s10971-011-2542-4

[37] Selbach SM, Einarsrud MA, Tybell T, Grande T. Synthesis of $BiFeO_3$ by wet chemical methods. Journal of the American Ceramic Society. 2007;**90**:3430-3434. DOI: 10.1111/j.1551-2916.2007.01937.x

[38] Xian T, Yang H, Shen X, Jiang JL, Wei ZQ, Feng WJ. Preparation of high-quality $BiFeO_3$ nanopowders via a polyacrylamide gel route. 2009;**480**:889-892

[39] Yang H, Xian T, Wei ZQ, Dai JF, Jiang JL, Feng WJ. Size-controlled synthesis of BiFeO3 nanoparticles by a soft-chemistry route. Journal of sol-gel science and technology. 2011;**58**:238-243. DOI: 10.1007/s10971-010-2383-6

[40] Liu J, Li M, Pei L, Wang J, Yu B, Wang X, et al. Structural and multiferroic properties of the Ce-doped $BiFeO_3$ thin films. Journal of Alloys and Compounds. 2010;**493**(1-2): 544-548. [Internet] Available from:. DOI: 10.1016/j.jallcom.2009.12.152

[41] Reda SM. Materials science in semiconductor processing synthesis of ZnO and $Fe_2O_3$ nanoparticles by sol-gel method and their application in dye-sensitized solar cells. Materials Science in Semiconductor Processing. 2011;**13**(5-6):417-425. [Internet] Available from:. DOI: 10.1016/j.mssp.2011.09.007

[42] Kempen PJ, Hitzman C, Sasportas LS, Gambhir SS, Sinclair R. Advanced Characterization Techniques for Nanoparticles for Cancer Research: Applications of SEM and NanoSIMS for Locating Au Nanoparticles in Cells. 2014:157-63.

[43] Goldstein A, Soroka Y, Frušić-Zlotkin M, Popov I, Kohen R. High

resolution SEM imaging of gold nanoparticles in cells and tissues. Journal of Microscopy. 2014;**256**(3):237-247

[44] Rades S, Hodoroaba V-D, Salge T, Wirth T, Lobera MP, Labrador RH, et al. High-resolution imaging with SEM/T-SEM, EDX, and single morphological and elemental analyses of single engineered nanoparticles. RSC Advances. 2014;**4**:49577-49587. DOI: 10.1039/c4ra05092d

[45] Borchert H, Shevchenko EV, Robert A, Mekis I, Kornowski A, Gru G, et al. Determination of nanocrystal sizes : A comparison of TEM, SAXS, and XRD studies of highly monodisperse $CoPt_3$ particles. Langmuir. 2005;**21**:1931-1936. DOI: 10.1021/la0477183

[46] Vilela D, González MC, Escarpa A. Sensing colorimetric approaches based on gold and silver nanoparticles aggregation : Chemical creativity behind the assay. A review. Analytica Chimica Acta. 2012;**751**:24-43. [Internet]. DOI: 10.1016/j.aca.2012.08.043

[47] Nam J, Won N, Jin H, Chung H, Kim S. pH-induced aggregation of gold nanoparticles for Photothermal cancer therapy. Journal of the American Chemical Society. 2009;**131**:13639-13645

[48] Perrault SD, Chan WCW. In vivo assembly of nanoparticle components to improve targeted cancer imaging. Proceedings of the National Academy of Sciences. USA. 2010;**107**:11194-11199. DOI: 10.1073/pnas.1001367107

[49] Mahl D, Diendorf J, Meyer-zaika W, Epple M. Colloids and surfaces a: Physicochemical and engineering aspects possibilities and limitations of different analytical methods for the size determination of a bimodal dispersion of metallic nanoparticles. Colloids and Surfaces A: Physicochemical and Engineering Aspects. 2011;**377**(1-3): 386-392. [Internet] Available from:. DOI: 10.1016/j.colsurfa.2011.01.031

[50] Courty A, Mermet A, Albouy PA, Duval E, Pileni MP. Vibrational coherence of self-organized silver nanocrystals in f.c.c. supra-crystals. Nature Materials. 2005;**4**:395-398. DOI: 10.1038/nmat1366

[51] Shevchenko EV, Talapin DV, Kotov NA, Brien SO, Murray CB. Structural diversity in binary nanoparticle superlattices. Nature. 2006;**439**(January):55-59. DOI: 10.1038/nature04414

[52] Chem JM, Singamaneni S, Bliznyuk VN, Tsymbal EY. Magnetic nanoparticles : Recent advances in synthesis, self-assembly, and applications. Journal of Materials Chemistry. 2011;**21**:16819-16845. DOI: 10.1039/c1jm11845e

[53] Yan W, Petkov V, Mahurin SM, Overbury SH, Dai S. Powder XRD analysis and catalysis characterization of ultra-small gold nanoparticles deposited on titania-modified SBA-15. 2005;**6**:404-408

[54] Mazzaglia A, Monsu L, Mezzi A, Kaciulis S, Caro T De, Ingo GM, et al. Supramolecular colloidal systems of gold nanoparticles/amphiphilic cyclodextrin : an FE-SEM and XPS investigation of nanostructures assembled onto solid surface. 2009;**113**:12772-12777. DOI: 10.1021/jp903673x

[55] Ju-nam Y, Lead JR. Manufactured nanoparticles : An overview of their chemistry, interactions, and potential environmental implications. Science Total Environment. 2008;**400**(1-3):396-414. [Internet] Available from:. DOI: 10.1016/j.scitotenv.2008.06.042

[56] Reddy LH, Arias JL, Nicolas J, Couvreur P. Magnetic nanoparticles:

design and characterization, toxicity and biocompatibility, pharmaceutical and biomedical applications. Chemical Reviews. 2012;**112**:5818-5878. DOI: 10.1021/cr300068p

[57] Hurley KR, Ring H, Kang H, Klein ND, Haynes CL. Characterization of magnetic nanoparticles in biological matrices. Analytical Chemistry. 2015;**87**: 11611-11619. DOI: 10.1021/acs. analchem.5b02229

[58] Delvallée A, Feltin N, Ducourtieux S, Trabelsi M, Hochepied JF. Direct comparison of AFM and SEM measurements on the same set of nanoparticles. Measurement Science and Technology. 2015;**26**:085601

[59] Russo R, Esposito E, Granata C, Vettoliere A, Russo M. Magnetic nanoparticle characterization using nano-SQUID based on niobium Dayem bridges. In: Physics Procedia. 2012;**36**:293-299. [Internet]. DOI: 10.1016/j.phpro.2012.06.162

[60] Granata C, Russo R, Esposito E, Vettoliere A, Russo M, Musinu A, et al. Magnetic properties of iron oxide nanoparticles investigated by nanoSQUIDs. The European Physical Journal B. 2013:2-6

[61] Kumari M, Widdrat M, Tompa É, Uebe R, Schüler D, Pósfai M, et al. Distinguishing magnetic particle size of iron oxide nanoparticles with first-order reversal curves. Journal of Applied Physics. 2014;**116**. DOI: 10.1063/1.4896481

[62] Upadhyay S, Parekh K, Pandey B. Influence of crystallite size on the magnetic properties of $Fe_3O_4$ nanoparticles. Journal of Alloys and Compounds. 2016;**678**:478-485. [Internet]. DOI: 10.1016/ j.jallcom.2016.03.279

[63] Li W, Zamani R, Rivera Gil P, Pelaz B, Ibáñez M, Cadavid D, Shavel A, Alvarez-Puebla RA, Parak WJ, Arbiol J, Cabot A. CuTe Nanocrystals: shape and size control, plasmonic properties, and use as SERS probes and photothermal agents. Journal of the American Chemical Society. 2013;**135**:7098-7101. DOI: 10.1021/ja401428e

[64] Thi N, Thanh K, Rosenzweig Z. Development of an aggregation-based immunoassay for anti-protein a using gold nanoparticles. Analytical Chemistry. 2002;**74**(7):1624-1628

[65] Liu J, Wu K, Wang J-P. Magnetic properties of cubic FeCo nanoparticles with anisotropic long-chain structure. AIP Advances. 2016;**6**(5):056126. [Internet] Available from:. DOI: 10.1063/ 1.4945042

[66] Friedrich T, Lang T, Rehberg I, Richter R. Spherical sample holders to improve the susceptibility measurement of superparamagnetic materials. Review of Scientific Instruments. 2012;**83**:1-7. DOI: 10.1063/1.3700185

[67] Lu LT. Department of Chemistry Water-dispersible Magnetic Nanoparticles for Biomedical applications: Synthesis and characterisation. 2011

Chapter 3

# Ballistic Composites, the Present and the Future

*Stevan Stupar*


## Abstract

In recent decades, the expansion of arms development has initiated the need to increase the protection of people and vehicles from pistol and rifle ammunition. Modern ceramic and ceramic-based materials are lightweight and durable and provide a sufficient level of protection against the penetration and impact of ammunition, which can protect the vital organs of the person. Modern tendencies require the addition of armor to vehicles, which reduces the necessity of excessive bulk steel usage and eliminates large and heavy mass weight amounts. By replacing the armored steel with new ballistic materials, a higher level of ballistic protection could be achieved, as well as reduction of weight, which both allows better mobility and increases the ability of installment of additional battle fighting equipment. Modern ceramic materials used in the production of armor are made by sintering the ceramic powder under certain conditions in a suitable molding tool. The chapter will cover the short material requirements, and material responses to ballistic impact, production methods, and applications. Also, the chapter will include the usage of ceramic fibers, alumina, silicon and boron carbide, titanium diboride, and ballistic materials that consist of a ceramic face bonded to a reinforced plastic laminate or metallic backplate.


**Keywords:** ballistic protection, lightweight composites, oxide ceramics, non-oxide ceramics, alumina, silicon carbide

## 1. Introduction

In many situations through history, it has been proven that saving every life is priceless and that no machine can replace its existence in active conflict and peace-keeping times. In modern warfare, it is important that the soldiers and high mobility vehicles are equipped with sufficient and effective ballistic protection so that used materials do not negatively affect the performance of tactical actions of an individual or a vehicle. For that purpose, depending on the mission, flexible body armor, molded breastplates and helmets could be used for the body protection of the individuals. Adequate ballistic plates are needed to be designed in a way to able to prevent projectiles of various calibers to penetrate, damage, and destroy armored vehicles such as transporters, tanks, helicopters .

This chapter contains the definition and classification of armor materials with a focus on ceramics and ceramic composites, the usage of ceramic fibers, processing of ceramics-based ballistic materials, and the design and manufacture of ceramic-based composite ballistic materials. The primary goal of this chapter is to explain to somebody outside this field of research and development the significance of the development and application of ceramics and composites based on ceramics in ballistic protection, production methods and secondary explain a problem that has always existed is how to describe to somebody term "Bulletproof".

## 2. Definition and classification of ballistic protection materials

The development of armaments throughout history had to be accompanied by the development of shields. From ancient times until today, there has been a constant race in the development of weapons and ballistic protection of personnel and vehicles. Since the metal age, the shields have been made of metal, which provided sufficient protection. In the middle of the twentieth century, research has begun on the possible usage of composite materials in protection against projectile penetration. The sudden turn in armor-making technology of that era looked like science fiction. Serial production of composite helmets began in the 1970s in the USA and Great Britain, followed by the production of composite-based armored panels for combat vehicles. Today, modern materials for ballistic protection of humans and vehicles must meet certain strict requirements. Such materials can be defined as materials that must be generally light, cost-effective, low density, high compressive strength, high hardness, durability, and capable of retaining or breaking a projectile penetrator of a certain caliber. Nowadays, ballistic materials can be divided into four basic categories:

- Metal (e.g., aluminum, armor steel),
- Polymers (e.g., polyethylene, aramid),
- Ceramics (e.g., alumina, silicon carbide, boron carbide, and titanium diboride),
- Composite of the above materials (e.g., tandem armor system).

As mentioned earlier, metals are the oldest materials used for defensive purposes to cover the body of combat operations. Modern aluminum-based ballistic materials are usually made of 7xxx series aluminum alloy. According to studies, the performance of these alloys can be cured by heat treatment of aging, and sintering can be controlled by particle size. In particular, 7039 aluminum alloys, due to their high strength and ability to absorb energy, are of exceptional importance. These materials are used in combat vehicles as armor material [1–5]. Armor plate of hardened steel has been used for many years to provide protection of objects against impact damage. Commercial representatives of steel used in ballistic vehicle protection are Mars ® 300, high-strength steels, namely AISI 4340 and DIN 100Cr6 [6, 7].

The polymeric-based materials used for reinforcement of matrix intended for ballistic protection of personnel and vehicle can be divided at the para-aramid group (e.g., Kevlar ® and Twaron®), ultrahigh-molecular-weight polyethylene (UHMWPE) (e.g., Spectra®, Dyneema®, and Technora®) and liquid-crystal polymer fibers (e.g., Zylon® and Vectran®) [8]. Polymeric materials are primarily lightweight, but the primary disadvantage of ballistic composites made of polymeric materials are sensitive to high temperature, humidity, radiation, ultraviolet (UV) light, etc., it is very important to study the durability and reliability of body armor.

There is a wide range of ceramics and ceramic-matrix composites that can be used in protection against projectile penetration. These ceramic materials can be divided into oxide ceramics (mostly, alumina ceramics with different contents of $Al_2O_3$) and non-oxide ceramics (mostly carbides, nitrides, borides, and their combinations) [9]. Alumina due to its high density (3.95 g $cm^{-3}$), relatively high physical properties, low cost, and easy production is the most widely used oxide ceramics used to make ballistic materials [10]. Other types of non-oxide ceramic materials are carbides (e.g., silicon carbide-SiC and boron carbide-B4C) nitrides (e.g., silicon nitride-SiN) and borides (e.g., titanium diboride-TiB2). Enumerated non-oxide ceramics are more expensive than aluminum and a combination of different non-oxide ceramics is also possible [11–13].

Armor systems made of ceramics and composite materials are widely used in ballistic applications to repel armored missiles using materials with substrate properties and materials that minimize breakage. In order to make a quality ballistic composite, it is necessary to find the optimal composition and conditions to make them.

## 3. Ceramics and ceramic-composite as a lightweight material for ballistic protection

Armor systems made of ceramics and composite materials are widely used in ballistic applications to repel armored missiles using materials with substrate properties and materials that minimize breakage. In order to make a quality ballistic composite that can meet the requirements of modern warfare, it is necessary to find the optimal composition and conditions for making the composite. The main objective of ballistic material is to develop protection systems that are both effective and lighter in weight. Ceramic and ceramic-composite are lightweight materials that can provide the level of armor protection as 5083 Al and high-hard steel, but their application reduces the mass of soldiers` equipment or the entire combat vehicle, which increases their mobility, unlike metal applications. Also, usage of these materials for body armor needs to purchase the trauma reduction caused via projectile impact.

### 3.1 Ceramic fibers for ballistics

The ceramic fibers possess excellent physical and mechanical properties (e.g., high-strength and high-modulus properties). Due to their high resistance to very high temperatures, these fibers had found usage in the aerospace and rocket industry in manufacturing objects able to sustain the high level of physical and mechanical load [14]. Because these fibers have a large diameter, they can be used as uni-directional tapes in the prepreg manufacturing process of clay-fibers composites. Characteristics of composites intended for ballistic protection largely depend on fiber type and layer orientation. The ceramic fibers layer tends to be strongest when the load acts along the direction in which the fibers are laid. The impact of the projectile causes the development of longitudinal and transverse waves, which help define the mode of failure [15]. Also, many kinds of research prove that the direction and structure of fibers have an effect on ballistic resistance properties. The development of ballistic resistance plates with different fibers and fabrics have been explored for performance. Woven fabrics were found to provide better mechanical properties than unidirectional fabrics [16].

The ceramic fibers are usually made from large-diameter monofilaments tungsten-core wire and vapors of ceramic materials (e.g., boron and silicon carbide) in the vapor deposition process, and spinning method to obtain alumina ceramic fibers [17]. The physical and mechanical properties of different ceramic fibers are shown in **Table 1**.

Unlike other fibers used in the manufacture of composites to increase the ballistic protection of soldiers and vehicles (poly aramids, glass, aromatic polyesters, and UHMWPE), ceramic fibers can withstand temperatures up to 12,000°C. The materials for ballistic protection do not need to withstand such high temperatures, and the primary need is to prevent penetration of projectile. Fabrics made of ceramic fibers for the purpose of making composite materials in order to increase ballistic protection can be two-dimensional (2D) and three-dimensional (3D) fabrics. Two-dimensional woven fabrics are mainly used in the production of composites for ballistic applications. In laminated ballistic composites, different types of yarns that are intertwined can be combined, i.e. different yarns that extend along the length of the fabric (warp) and yarns that go from edge to edge (weft) based on a predefined pattern. The combination of layers that can be seen in one or more directions improves ballistic resistance and puncture resistance, resulting in multiple 2D woven fabrics. Multiple 2D woven fabrics can be layered to provide ballistic and puncture resistance, in particular by enhancing the ballistic and stabbing resistance, especially by decreasing the back-face deformation. The disadvantage of 2D woven fabric is that there is a high possibility of sequential delamination due to projectile impact and weakening of the adhesion caused by the deterioration of the matrix [18, 19]. The presence of Z-oriented fibers in 3D enhances in-plane properties due to the bias yarn layers so that could be the solution for the delamination problem. **Figure 1** is shown the various weave constructions of 2D and 3D fabrics used in ballistic composite production.

**Table 2** shows the density and Hugoniot elastic limits (HEL) of different ceramic fibers which can be used in the production process of ballistic protection products.

From the aspect of economic profitability, the use of alumina ceramic fibers is the most favorable among advanced ceramics fibers with high physical and mechanical properties. However, by the analysis shown in **Table 2**, using alumina (Nextel 3 M) has the lowest projectile penetration protection efficiency of all the materials shown but composites made on the basis of these fibers have the lowest efficiency of all the listed materials. Carbides are the hardest ceramics but do not withstand high impact pressures due to an amortization process that weakens the ceramic [20]. The Beryllium-oxide and Magnesia.

| Type | Density (g $cm^{-3}$) | Elastic modulus (GPa) | Tensile strength (MPa) | Strain to Failure (%) |
|---|---|---|---|---|
| Alumina (Nextel 3 M) | 2.50 | 152 | 1720 | 2.1 |
| Silicon Carbide (Nippon) | 2.80 | 418 | 4000 | 0.7 |

**Table 1.**
*The physical and mechanical properties of alumina and SiC fibers.*

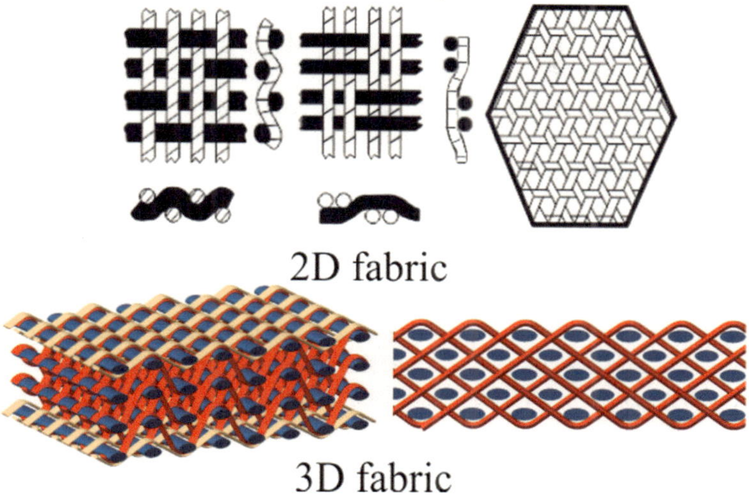


**Figure 1.**
*Weave constructions of 2D and 3D fabrics.*

| Type | Density (g cm$^{-3}$) | HEL |
|---|---|---|
| Alumina ($Al_2O_3$) | 2.50 | 11.2 |
| Boron Carbide ($B_4C$) | 2.82 | 15.0 |
| Beryllium-oxide (BeO) | 2.84 | 8.5 |
| Magnesia (MgO) | 3.57 | 8.9 |

**Table 2.**
*The density and Hugoniot elastic limits (HEL) of different ceramic fibers.*

### 3.2 Aluminum-oxide: alumina ($Al_2O_3$)

It was mentioned earlier that due to their good physical and mechanical properties, they are the most often used in the ballistic protection of soldiers and vehicles. Alumina provides excellent impact resistance, chemical resistance, abrasion resistance, and high-temperature properties and this material is cheaper than other ceramic materials for this purpose. Alumina can be found in several different phases: alpha ($\alpha$), beta ($\beta$), gamma ($\gamma$), eta ($\eta$), qi ($\chi$), delta ($\delta$) and cap ($\kappa$) alumina [21–25]. The $\alpha$-alumina particles look like white powder, have a small active surface, are more resistant to high temperatures, and have very good physical and mechanical properties. They are most often used as ceramic materials. $\beta$-alumina has a hexagonal structure with a lamellar structure. The $\gamma$-alumina particles are also nano-aluminum oxides of high purity, and also have a large active surface area. Lattices of this type of material are porous and stable at high temperatures. By modifying the structure of $\gamma$-alumina, they can be used as an adsorbent and/or catalyst [26]. $\eta$-$Al_2O_3$ particles have a similar crystal structure as $\gamma$-$Al_2O_3$ and a large specific surface area (2200 m$^2$ g$^{-1}$), which is why they can be used without modification as adsorbents or as catalysts if certain modifications are made [27]. Alumina in the $\chi$-$Al_2O_3$ phase is a metastable material of hexagonal crystal structure that has high thermal stability and the ability to bind metal cations, which is why it can be used as a catalyst [28]. The structure of δ-$Al_2O_3$ has been studied for over fifty years, and recent research has shown that at this stage there is a complex structure created by the internal development of various polytypes from tetrahedral to octahedral structure [29]. Particles κ-$Al_2O_3$ represent one of the transition phases of aluminum oxide, which has a polyhedral crystal structure in which oxygen and aluminum usually form octahedra and tetrahedra [30]. The chemical routes of alumina powder synthesis for sintering of ballistic protective equipment can be obtained by sol–gel [31], control precipitation [32], and hydrothermal processing [33] methods.

The mass fraction of alumina in ballistic plates is generally from 96 to 99%. The following **Figure 2** shows the morphology of alumina ballistic plates. **Table 3** shows the physical and mechanical properties of alumina-based ballistic plates manufactured by CeramTec12000 (https://www.ceramtec-industrial.com/en/ballistic-protective-ceramics).

According to the properties shown in **Table 3**, it can be concluded that with an increase in the mass fraction of $Al_2O_3$, the improvement of physical and mechanical properties is accompanied by an increase in density and mass. Impurities will mostly come from magnesium-oxide added to be avoided uncontrolled grain growth and enable the increased ballistic efficiency of corundum.

### 3.3 Silicon carbide (SiC)

Aside from alumina materials which are widely in use today, silicon carbide will be used where significant weight reduction or increased mechanical properties are

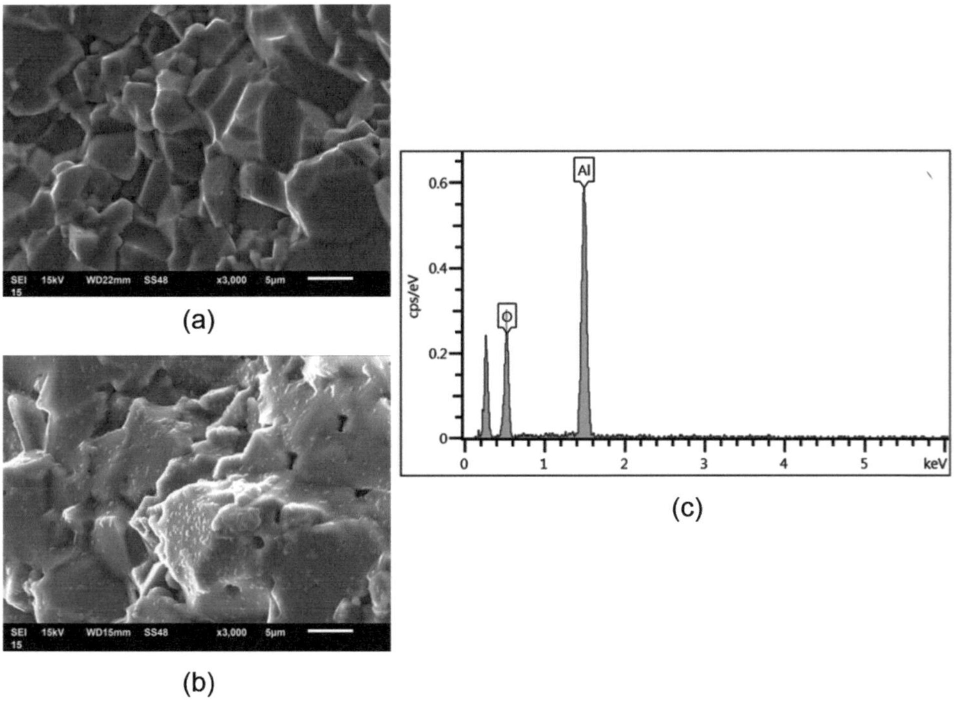


**Figure 2.**
*SEM photographs of alumina alumina-based ballistic plates with 96 (a) and 99 (b) mass % and characteristic EDS spectra.*

| | | 96% $Al_2O_3$ | 98% $Al_2O_3$ | 99% $Al_2O_3$ |
|---|---|---|---|---|
| Density | g $cm^{-3}$ | 3.75 | 3.80 | 3.87 |
| Residual porosity | % | <2 | <2 | <2 |
| Medium Grain Size | μm | 5 | 6 | 10 |
| Vickers Hardness | GPa | 12.5 | 13.5 | 15 |
| Young's Modulus | GPa | 310 | 335 | 365 |
| Bending Strength | MPa | 250 | 260 | 280 |
| Fracture Toughness | MPam 0.5 | 3.5 | 3.5 | 3.6 |

**Table 3.**
*Physical and mechanical properties of alumina-based ballistic plates manufactured by CeramTec.*

required. This ceramic material has very high hardness and a high strength-to-weight ratio, which makes Silicon Carbide very convenient for armor applications. Besides mentioned properties, the significant parameter for ballistic protection usage is the high impact resistance of the material. The physical and mechanical properties depend on the method of production. It can be produced by Sintering, hot-pressing, hot-isostatic pressing, reaction bonding, and spark plasma sintering. The processing temperature of Silicon Carbide is approximately 1600°C. Nowadays, studies of silicon carbide properties have shown that it was found that one-dimensional nanostructures such as wires, rods, and tubes have received continued interest as a result of their excellent physical and mechanical properties, in comparison to their bulk counterparts. **Figure 3** shows the morphology and EDS spectra of Silicon Carbide ballistic plate and at **Table 4** the physical and mechanical properties of Silicon Carbide-based ballistic plates manufactured by CeramTec.

The study in which they have examined the ceramic composite plate based on Silicon Carbide and Dyneema fibers provide ballistic protection. This type of composite which combines Silicon Carbide and Dyneema fibers provides physical and mechanical properties ballistic protection to the US National Institute of Justice level four (NIJ IV) standards indicating that the plate can resist .30 cal steel core armor-piercing rifle ammunition [34].

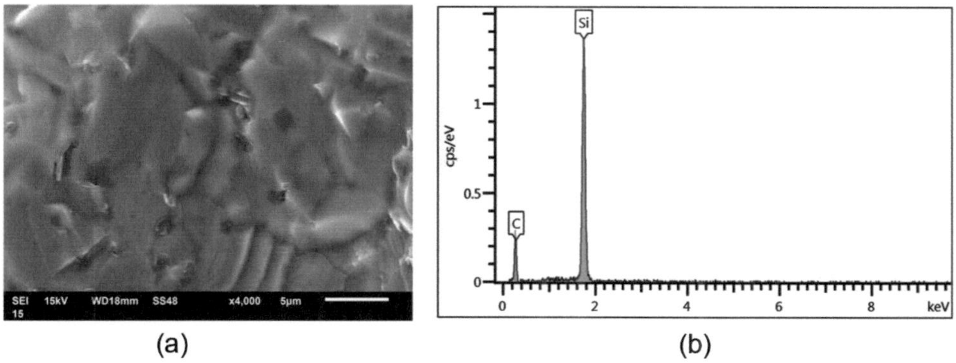


**Figure 3.**
*SSEM photograph (a) and characteristic EDS spectra (b) of silicon carbide ballistic plate.*

| | | SiC |
|---|---|---|
| Density | $g\ cm^{-3}$ | 3.1 |
| Residual porosity | % | <2.5 |
| Medium Grain Size | Mm | <5 |
| Vickers Hardness | GPa | 26 |
| Young's Modulus | GPa | 410 |
| Bending Strength | MPa | 400 |
| Fracture Toughness | MPam 0.5 | 3.2 |

**Table 4.**
*Physical and mechanical properties of SiC ballistic materials manufactured by CeramTec.*

### 3.4 Boron carbide ($B_4C$)

Boron Carbide is a ceramic material with outstanding physical and mechanical properties such as high hardness Hugoniot elastic limit and low specific weight which makes this material good for lightweight ballistic protection. The testing of this ceramic material pointed out a very interesting parameter, which is the high value of HEL (15–20 GPa). The material exhibits an abrupt drop in strength at very high pressures and impact rates. The production of light materials for ballistic protection based on boron carbide 95–99% of the theoretical density and grain size in the range between 1.5 and 60 mm is usually done by hot pressing and sintering without pressure [35]. **Table 5** shows the physical and mechanical properties of Boron Carbide lightweight ballistic material.

| | | $B_4C$ | |
|---|---|---|---|
| Density | $g\ cm^{-3}$ | 2.52 | [36] |
| Compressive strength | GPa | 3–5 | [37] |
| Flexural strength | MPa | 350–500 | [36] |
| High hardness | GPa | 25–30 | [38] |
| Young's Modulus | GPa | 450–550 | [39] |
| Fracture Toughness | MPam 0.5 | 2.7–3.6 | [35] |

**Table 5.**
*Physical and mechanical properties of $B_4C$ ballistic materials.*

The physical and mechanical properties of $B_4C$ ballistic materials largely depend on the method of processing and raw material. The ballistic characteristic of ceramic material based on boron carbide can be reinforced by the addition of silicone or titanium to obtain better ballistic performance. Also, raw $B_4C$ matrix can be reinforced with the addition of titanium-diboride powder to produce a composite with better ballistic performance [40]. From the point of fabrication, the composite based on Boron Carbide modified by the addition of Silicon Carbide is easier to form into the complex shapes needed for torso body armor and is nearly as effective ballistically as boron carbide itself.

### 3.5 Titanium-diboride ($TiB_2$)

Titanium-diboride with a melting point of 3225°C is classified as ultra-high-temperature ceramics. The monolithic $TiB_2$ and $TiB_2$-based ceramic composites are materials that, due to their physical and mechanical properties (e.g., HEL, density, strength, toughness, and hardness, etc.), have wide application, but can most often be used in cutting tools and materials for ballistic protection [41–43]. Since this material is fireproof, it can also be used at high temperatures, but it has been determined that at temperatures higher than 1400°C, the mechanism of plastic deformation is activated, which leads to grain-boundary sliding and creep boundary [43]. In order to increase the quality of ceramic products based on $TiB_2$, a series of tests of microstructural design control procedures have been initiated. Thus, research in recent years has led to improvements in the sintering process compared to traditional non-pressure techniques that are more efficient in terms of economic viability and physical-mechanical properties of the product. Advanced methods such as external pressure contribute to the consolidation of the ceramic structure and obtaining a dense mass. Hot pressing or hot isostatic by pressing have enabled the improvement of the microstructure of fireproof materials with reduced processing temperatures. In recent years, the Spark Plasma Sintering process has played a significant role in the production of ceramic materials, in which the density of the final product is increased by pulsed electrical flow electricity for a shorter period of treatment. **Table 6** shows the physical, mechanical, and oxidation properties of Titanium-diboride [44].

By comparison of physical and mechanical properties in **Table 5** and previous one showed properties of Alumina, Silicon Carbide and Boron Carbide it can be concluded that is all parameters higher in the case of Titanium-diboride than previously explained ceramics materials. In the case of the production and application of composites based on titanium-diboride, due to the increased density, the final product would have a higher mass compared to previous ceramics. Therefore, optimal parameters should be found in the production of composites.

### 3.6 Ceramics-based ballistic protection composites

The constant race in armaments and the development of new types of weapons with different calibers, requires a response in protection from the same. Therefore, it is necessary to constantly develop ballistic materials that would protect people and vehicles on duty. Today, modern armies' function on the basis of three postulates: do not notice, do not shoot, and do not break through armor. The last, perhaps the most important requirement depends on the material. The efficiency of existing materials can be improved with ceramic-based composites.

The ceramic material can be reinforced by metal/metal oxide powder addition before sintering. This method of ceramic reinforcement was described by A.

| | | $TiB_2$ |
|---|---|---|
| Density | g cm$^{-3}$ | 4.5 |
| Elastic modulus | GPa | 560 |
| Flexural strength | MPa | 700–1000 |
| High hardness | GPa | 25–35 |
| Fracture Toughness | MPam 0.5 | 4–5 |
| Oxidation resistance | °C | <1200 |

**Table 6.**
*Physical, mechanical, and oxidation properties of $TiB_2$.*

Pettersson et al. in the study of titanium-diboride reinforced by titanium powder addition before the sintering process [13]. The precursor powders were mixed in a high-speed planetary mil and after drying the matrix was treated by spark plasma sintering. The results of mechanical tests show that the composite with titanium contains between 5 and 6 wt. % have the best physical and mechanical properties. In the case of fracture toughness tests, this parameter is less sensitive to titanium content, except when 10% by weight is used, where toughness begins to increase rapidly. At the same procedure, ceramic composite can be reinforced by the addition of aluminum and magnesium [45], zirconia [46], and silicon [47].

Different reinforcement method of ceramics is the addition of ultra-high molecular weight polyethylene or low-density polyethylene (LDPE) before sintering by the hot-pressing process to fill possible cavities (pores) formed during sintering. Oliveira et al. [48] had examined the ballistic properties of alumina-based composites reinforced by the different mass fractions of low-density polyethylene. In the stated examination, alumina-based composites were modified by 10–30 wt. % of low-density polyethylene. Results obtained by these tests lead to the following conclusions: LPDE has the function of keeping the alumina grains cohesive, LDPE among the alumina particles helps to increase the toughness of the composite and optimal addition of 20 wt. % of LDPE.

Ballistic ceramic-based composites can also be reinforced by a metal plate. The bonding of different materials by adhesion allows the composite to withstand the acoustic waves generated by the projectile impact. Due to the existence of acoustic impedance, in such composites, the effect of matching impedances produced by acoustic waves during projectile impact is reduced. Reducing this effect improves the physical and mechanical properties and prevents cracking of the ceramic armor, which allows one armor plate to protect itself from multiple impacts, provides structural integrity, and, to a lesser extent, the efficiency of the ballistic mass of the composite ceramic armor. Due to their increased mass, such composites are not used to increase the ballistic protection of people, but mainly vehicles. Due to the specific mass of metal and the load of the object, which requires an increase in ballistic protection, aluminum is most often used to strengthen ceramics-based composites. The main challenge in producing these composites is choosing the right adhesive. The adhesive must provide a ceramic-metal bond, not to allow easy separation of the composite during the impact of the projectile and to withstand the force of acoustic waves. The main criteria for choosing an adhesive is the ability to keep composites bonded after an impact of the projectile. The service life of the composite also depends on the quality of the adhesive. For this purpose, polymeric and high-temperature adhesives are used for ceramic-based composites production. One approach to improving the ballistic performance of ceramic armor with metal backings is to create a strong metallurgical bond and functionally graded composites in which metal layers transition to the ceramic layers without damaging interfaces. In ceramic-based composites, aluminum as backing material can be changed by other ceramics, Kevlar®, Twaron®, fiberglass, UHMWPE, and polyamide which can undergo thermal treatment of the glued ceramic product with backing material in an autoclave where temperature, pressure, and vacuum are applied.

## 4. Ceramic-based ballistic materials-processing and equipment

The physical and mechanical properties of ceramic-based ballistic materials largely depend on the processing methods and tooling. The density, high-strength, and high-modulus properties are influenced by the processing method, and the processing method is influenced by a number of factors. The factors that determine the quality of the material are the size and shape of raw materials, availability of tooling for that particular size and shape, compatibility of different raw materials and adhesives, resin content, economical parameters (cost of raw materials and labor cost). The secondary factor on making the selection in the manufacturing process depends on the main target which protects as well. This factor depends on the ballistic threat, the maximum acceptable weight for personnel and/or vehicles, structural and impact requirements, exposure to chemicals, fuels and lubricants, moisture, and other threats.

The most widespread processes for the production of ballistic materials based on ceramics are a hand-layup method, vacuum bagging, vacuum and oven processing, compression molding, and autoclave method.

### 4.1 Hand-layup method

Hand-layup method is also known as the wet-layup method. This process is the simplest process for the production of ballistic protection composites and is based on the application of materials in layers. The production of composites by this method is a combination of a polymer matrix reinforced mainly with fibers. The shape of the product depends on the mold used in which the layers of reinforcing fibers are placed, which are then saturated with wet (resin) by pouring over the fabric and pulling into the fabric. The resin is generally applied to reinforcing fibers or fabric with a roller. After applying the reinforcement to the resin matrix, the tool is left to harden the resin at room temperature or elevated depending on the type of matrix. The hand-layup method is suitable for the production of smaller quantities and in the conditions of manufacture and acceptable for prototype production in which complex molding or other costs might be an issue. The hand-layup method is not suitable for high-volume applications and has high labor costs.

### 4.2 Vacuum bagging

Vacuum bagging is an extension of the previously explained process by applying pressure to the laminate when it is laid to improve its consolidation. In this process, sealing of reinforcing fibers or fabrics is achieved. The air under the bag is extracted by means of a vacuum pump and in this way up to 1 atm of pressure can be applied to the laminate to solidify it. The main advantages of this process are higher fiber content laminates can usually be achieved than with standard hand-layup method, lower void content is achieved and better fiber wet-out due to pressure and resin flow throughout structural fibers, with excess into bagging materials, and the vacuum reduces the amount of volatiles emitted during cure. The disadvantages of the explained process are extra process costs, higher level of skilled operators needed, especially in mixing and control of resin content.

### 4.3 Vacuum and oven processing

Processing in a vacuum and in the oven is the processing of pre-impregnated fabrics and fabrics soaked in resin film is performed at atmospheric pressure and at elevated temperature. The vacuum bag surrounds the composite which binds the hermetic membrane and uses the ambient atmospheric pressure to compress the product in the vacuum bag, and, at the same time, heat is applied to the composite structure so that it takes the shape of a mold during the curing process. For all resin content used for the matrix is accurately set by the manufacturer. Using this method, higher fiber contents can be achieved. The disadvantage is higher material cost than previously described methods. The processing cost can be high, because of the autoclave required to cure the composite using high temperature and pressure.

## 4.4 Compression molding

Compression molding is the production of ballistic composites the heating of composite placed in mold cavity (female) or form (male) using a two-part mold system. The composite raw materials suffer applied force and pressure during contact with all mold areas, while heat and pressure are maintained until the molding material has been cured or sintered. This process can be used for ceramic-plastic or ceramic-metal composite production and heating of thermosetting resins for adhesives for the curing process. With compression on one side only, the result is an increase in pressure on the other side equal to the amount of vacuum being generated. Compression molding produces high-strength composite structures and complex parts in a wide variety of sizes. Compression-molded fiberglass or composite parts are characterized by net size and shape, two excellent finished surfaces, and outstanding part-to-part repeatability. This type of method enables the production of large-size parts beyond the capacity of extrusion techniques, and it can apply to composite thermoplastics with ceramic materials, UD tapes, woven fabrics, randomly oriented fiber mats, or chopped strands, also ceramics can be sintered by this method. Compression molding is one of the lowest cost options for the molding of complex composite parts. However, the slow production and limited largely too flat or moderately curved parts with no undercuts are disadvantages of this method.

## 4.5 Autoclave processing

Autoclave processing is an advanced pressure-bag and the vacuum-bag molding process, and represents the production of composite materials uses denser molds without cavities because higher heat and pressure are used for curing. Autoclaves are essentially heated pressure vessels usually equipped with vacuum systems in which the packed layer on the mold is taken for the curing cycle. This method has wide usage, in the aerospace industry to fabricate high strength-to-weight ratio parts from impregnated high-strength fibers for aircraft, spacecraft, and missiles. The heating mechanism in the autoclave can be direct or indirect. Indirect heating systems have the heat source outside the autoclave and transfer heat into the working envelope by means of a heat exchanger. Direct heating systems have their heat source within the autoclave and aim to maximize the heat transfer from the elements to the pressure medium.

## 4.6 Sintering processes for ceramic production

The production of ballistic ceramic production requires far higher temperatures than for crosslinking polymeric materials used in the manufacture of ballistic composites. At high sintering temperatures of ceramic materials, none of the polymeric materials such as Kevlar®, Twaron®, and others can resist. Therefore, ceramic materials must be specially made of a backing made of plastic. According to new research, ceramic and/or glass fibers that can withstand high sintering temperatures can be added to ceramic materials before sintering.

In accordance with the pressure amount applied, sintering can be divided into pressureless and pressurized sintering. Further, pressureless sintering can be divided into reaction sintering, thermal plasma sintering, microwave sintering, and atmospheric sintering. The second group is pressurized sintering which can be divided into solid (hot pressing, spark plasma sintering, and ultra-high-pressure sintering) and gas compaction (hot isostatic pressing and high-pressure gas reaction sintering).

The methods that represent pressureless sintering are usually economically acceptable methods for fabrication of ballistic ceramics than pressurized sintering, but these methods require the use of a high sintering temperature (≥1200°C) and the process least longer than 120 min for densification and solute homogenization. Nowadays, ballistic ceramics are mostly produced by spark plasma sintering [20, 43, 49] and hot-pressing sintering [37, 45].

Spark plasma sintering is a pressurized sintering method assisted by the pulsed-direct current process. The powder of raw material is loaded in an electrically conducting matrix and sintered under uniaxial pressure. Bypassing direct current through a matrix and sintered powder, if they are conductive, they act as a heating source, so in addition to external heating, internal heating is provided, which leads to improved heat transfer and rapid consolidation during sintering, which speeds up the production of ceramic materials for ballistic composites. This method can be used to produce large plates that can be machined later. Hot pressing is the most commonly used technique for fabricating dense, non-oxide monolithic ceramics and their composites. The ceramic materials can be produced in a mixture matrix (coil and graphite) and metal or metal-oxide powder under a uniaxial pressure hot-pressing but in the absence of direct current. At maximum pressure, the contact points between raw materials develop very high stress, increasing the local diffusion rates.

The time of processing, heating, densification, particle size, temperature, pressure, heating rate, and holding time all influence the physical and mechanical properties of the hot-pressed ceramics whilst a controlled atmosphere is required for the non-oxides. Carbides and borides are often hot-pressed under a vacuum or an inert gas such as argon whilst the nitrides are generally densified under a nitrogen atmosphere. Often, the pressure is applied when the maximum sintering temperature is reached, though it can be increased at intervals as the temperature increases.

## 5. Conclusion

The chapter Ballistic Composites, The Present and The future have attempted to present usage of ceramic materials and ceramic-based composites intended for needed for personnel and combat vehicles ballistic protection from penetration and impact of the projectile. Ballistic materials during the history of war follow the armament development to protect the soldier in the first place. Nowadays, metals and steels are increasingly being replaced by lightweight materials such as ceramics and composites, and in the future with the development of technology, metals and steels are likely to be completely phased out for ballistic protection. Ceramic materials such as alumina, SiC, $B_4C$, and $TiB_2$, have acceptable physical and mechanical properties to absorb and minimize breakage and stop the projectile penetration before reaching the vital organs and purchase the trauma reduction of the soldier or essential parts of the vehicle mechanism.

Ceramics and ceramic-based bulletproof materials can be divided into oxide and non-oxide ceramics materials and they are generally light, cost-effective, low density, high compressive strength, high hardness, durability, and capable of retaining or breaking a projectile penetrator of a certain caliber and fully can change the armor plate of aluminum or hardened steel. The main objective of ballistic material is to develop protection systems that are both effective and lighter in weight. The physical and mechanical properties of ceramic and ceramic-based composites depend on processing. The ballistic composite production methods which can be used are a hand-layup method, vacuum bagging, vacuum and oven processing, compression molding, and autoclave processing. Also, ceramic materials require higher temperatures and pressures than ceramic-plastic composite, and these materials can be produced by sintering processes. In this chapter, special attention was paid to processes of spark plasma sintering and hot-pressing methods.

According to the current tendency of weapons development and the requirements of modern battlefields, the tactics are primarily based on greater mobility of units while maintaining protection against the penetration of projectiles. Therefore, research and future production have turned to lightweight materials such as ceramic materials and their composites.

## Conflict of interest

The author declares no conflict of interest.

## Author details

Stevan Stupar[1,2]

1 Military Technical Institute, Belgrade, Serbia

2 Military Academy of the University of Defence, Belgrade, Serbia

*Address all correspondence to: stevan.stupar13@gmail.com

## References

[1] Dixit M, Mishra RS, Sankaran KK. Structure-property correlations in Al 7050 and Al 7055 high-strength aluminum alloys. Materials Science and Engineering A. 2008;**478**:163-172. DOI: 10.1016/j.msea.2007.05.116

[2] Ludtka GM, Laughlin DE. The influence of microstructure and strength on the fracture mode and toughness of 7XXX series aluminum alloys. Metallurgical Transactions A. 1982;**13**: 411-425. DOI: 10.1007/BF02643350

[3] Erdem M, Cinici H, Gokmen U, Karakoc H, Turker M. Mechanical and ballistic properties of powder metal 7039 aluminium alloy joined by friction stir welding. Transactions of Nonferrous Metals Society of China. 2016;**26**:74-84. DOI: 10.1016/S1003-6326(16)64090-6

[4] Dumont D, Deschamps A, Brechet Y. On the relationship between microstructure, strength and toughness in AA7050 aluminum alloy. Materials Science and Engineering A. 2003;**356**:326-336. DOI: 10.1016/S0921-5093(03)00145-X

[5] Zhao PZ, Tsuchida T. Effect of fabrication conditions and Cr, Zr contents on the grain structure of 7075 and 6061 aluminum alloys. Materials Science and Engineering A. 2009;**499**:78-82. DOI: 10.1016/j.msea.2007.09.094

[6] Fras T, Roth CC, Mohr D. Fracture of high-strength armor steel under impact loading. International Journal of Impact Engineering. 2018;**111**:147-164. DOI: 10.1016/j.ijimpeng.2017.09.009

[7] Demir T, Übeyli M, Yildirum RO. Effect of hardness on the ballistic impact behavior of high-strength steels against 7.62-mm Armor piercing projectiles. Journal of Materials Engineering and Performance. 2009;**18**:145-153. DOI: 10.1007/s11665-008-9288-3

[8] Ray BC. Impact of environmental and experimental parameters on FRP composites. In: Eighteenth Int. Symp. Process. Fabr. Adv. Mater. [PFAM XVIII]. Japan Society for the Promotion of Science; 2009. pp. 1-10

[9] Medvedovski E. Ballistic performance of Armour ceramics: Influence of design and structure. Part 2. Ceramics International. 2010;**36**:2117-2127. DOI: 10.1016/j.ceramint.2010.05.022

[10] Medvedovski E. Ballistic performance of Armour ceramics: Influence of design and structure. Part 1. Ceramics International. 2010;**36**:2103-2115. DOI: 10.1016/j.ceramint.2010.05.021

[11] Miyazaki H, Hyuga H, Yoshizawa YI, Hirao K, Ohji T. Relationship between fracture toughness determined by surface crack in flexure and fracture resistance measured by indentation fracture for silicon nitride ceramics with various microstructures. Ceramics International. 2009;**35**:493-501. DOI: 10.1016/j.ceramint.2008.01.006

[12] Boldin MS, Berendeev NN, Melekhin NV, Popov AA, Nokhrin AV, Chuvildeev VN. Review of ballistic performance of alumina: Comparison of alumina with silicon carbide and boron carbide. Ceramics International. 2021;**47**:25201-25213. DOI: 10.1016/j.ceramint.2021.06.066

[13] Pettersson A, Magnusson P, Lundberg P, Nygren M. Titanium-titanium diboride composites as part of a gradient Armour material. International Journal of Impact Engineering. 2005;**32**:387-399. DOI: 10.1016/j.ijimpeng.2005.04.003

[14] Yadav S, Ravichandran G. Penetration resistance of laminated ceramic/polymer structures. International Journal of Impact Engineering. 2003;**28**:557-574. DOI: 10.1016/S0734-743X(02)00122-7

[15] Cheeseman BA, Bogetti TA. Ballistic impact into fabric and compliant composite laminates. Composite Structures. 2003;**61**:161-173. DOI: 10.1016/S0263-8223(03)00029-1

[16] Chen X, Zhou Y, Wells G. Numerical and experimental investigations into ballistic performance of hybrid fabric panels. Composites. Part B, Engineering. 2014;**58**:35-42. DOI: 10.1016/j.compositesb.2013.10.019

[17] Bilisik K. Two-dimensional (2D) fabrics and three-dimensional (3D) preforms for ballistic and stabbing protection: A review. Textile Research Journal. 2017;**87**:2275-2304. DOI: 10.1177/0040517516669075

[18] Bilisik K, Korkmaz M. Multilayered and multidirectionally-stitched aramid woven fabric structures: Experimental characterization of ballistic performance by considering the yarn pull-out test. Textile Research Journal. 2010;**80**:1697-1720. DOI: 10.1177/0040517510365954

[19] Kadir Bilisik A, Turhan Y. Multidirectional stitched layered aramid woven fabric structures and their experimental characterization of ballistic performance. Textile Research Journal. 2009;**79**:1331-1343. DOI: 10.1177/0040517509104275

[20] Rahbek DB, Johnsen BB. Dynamic Behaviour of Ceramic Armour Systems. Norwegian Defence Research Establishment: Kjeller; 2015

[21] Hao Z, Wu B, Wu T. Preparation of alumina ceramic by κ-Al2O3. Ceramics International. 2018;**44**:7963-7966. DOI: 10.1016/j.ceramint.2018.01.235

[22] Favaro L, Boumaza A, Roy P, Lédion J, Sattonnay G, Brubach B, et al. Experimental and ab initio infrared study of χ-, κ- and α-aluminas formed from gibbsite. Journal of Solid State Chemistry. 2010;**183**(4):901-908. DOI: 10.1016/j.jssc.2010.02.010

[23] Pradhan JK, Bhattacharya IN, Das SC, Das RP, Panda RK. Characterisation of fine polycrystals of metastable η-alumina obtained through a wet chemical precursor synthesis. Materials Science and Engineering: B. 2000;**77**:185-192. DOI: 10.1016/S0921-5107(00)00486-4

[24] Esteban-Cubillo A, Díaz C, Fernández A, Díaz LA, Pecharromán C, Torrecillas R, et al. Silver nanoparticles supported on α-, η- and δ-alumina. Journal of the European Ceramic Society. 2006;**26**:1-7. DOI: 10.1016/j.jeurceramsoc.2004.10.029

[25] Yashnik SA, Kuznetsov VV, Ismagilov ZR. Effect of χ-alumina addition on H2S oxidation properties of pure and modified γ-alumina, Cuihua Xuebao/Chinese. Journal of Catalysis. 2018;**39**:258-274. DOI: 10.1016/S1872-2067(18)63016-5

[26] Paranjpe KY. Alpha, Beta and Gamma alumina as a catalyst - A review. The Pharma Innovation Journal. 2017;**6**:236-238

[27] Tilley DB, Eggleton RA. The natural occurrence of ETA-alumina (η-Al2O3) in bauxite. Clays and Clay Minerals. 1996;**44**:658-664. DOI: 10.1346/CCMN.1996.0440508

[28] Meephoka C, Chaisuk C, Samparnpiboon P, Praserthdam P. Effect of phase composition between nano γ- and χ-Al2O3 on Pt/Al2O3 catalyst in CO

oxidation. Catalysis Communications. 2008;**9**:546-550. DOI: 10.1016/j.catcom.2007.04.016

[29] Kovarik L, Bowden M, Genc A, Szanyi J, Peden CHF, Kwak JH. Structure of δ-alumina: Toward the atomic level understanding of transition alumina phases. Journal of Physical Chemistry C. 2014;**118**:18051-18058. DOI: 10.1021/jp500051j

[30] Ollivier B. Crystal structure of κ-alumina: An X-ray powder diffraction, TEM and NMR study. Journal of Materials Chemistry. 1997;**7**:1049-1056. DOI: 10.1039/a700054e

[31] Stupar SS, Vuksanović MM, Totovski LM, Jančić Heinemann RM, Mijin D. Adsorption of anthraquinone dye ab111 from aqueous solution using synthesized alumina-iron oxide doped particles. Science of Sintering. 2021;**53**:91-117. DOI: 10.2298/SOS2101091S

[32] Parida KM, Pradhan AC, Das J, Sahu N. Synthesis and characterization of nano-sized porous gamma-alumina by control precipitation method. Materials Chemistry and Physics. 2009;**113**:244-248. DOI: 10.1016/j.matchemphys.2008.07.076

[33] Lee HC, Kim HJ, Rhee CH, Lee KH, Lee JS, Chung SH. Synthesis of nanostructured γ-alumina with a cationic surfactant and controlled amounts of water. Microporous and Mesoporous Materials. 2005;**79**:61-68. DOI: 10.1016/j.micromeso.2004.10.021

[34] Wu KK, Chen YL, Yeh JN, Chen WL, Lin CS. Ballistic impact performance of SiC ceramic-Dyneema Fiber composite materials. Advances in Materials Science and Engineering. 2020;**2020**:477-483. DOI: 10.1155/2020/9457489

[35] Ghosh D, Subhash G, Sudarshan TS, Radhakrishnan R, Gao XL. Dynamic indentation response of fine-grained boron carbide. Journal of the American Ceramic Society. 2007;**90**:1850-1857. DOI: 10.1111/j.1551-2916.2007.01652.x

[36] Roy TK, Subramanian C, Suri AK. Pressureless sintering of boron carbide. Ceramics International. 2006;**32**:227-233. DOI: 10.1016/j.ceramint.2005.02.008

[37] Paliwal B, Ramesh KT. Effect of crack growth dynamics on the rate-sensitive behavior of hot-pressed boron carbide. Scripta Materialia. 2007;**57**:481-484. DOI: 10.1016/j.scriptamat.2007.05.028

[38] Vargas-Gonzalez L, Speyer RF, Campbell J. Flexural strength, fracture toughness, and hardness of silicon carbide and boron carbide armor ceramics. International Journal of Applied Ceramic Technology. 2010;7:643-651. DOI: 10.1111/j.1744-7402.2010.02501.x

[39] Domnich V, Reynaud S, Haber RA, Chhowalla M. Boron carbide: Structure, properties, and stability under stress. Journal of the American Ceramic Society. 2011;**94**:3605-3628. DOI: 10.1111/j.1551-2916.2011.04865.x

[40] Lo C, Li H, Toussaint G, Hogan JD. On the evaluation of mechanical properties and ballistic performance of two variants of boron carbide. International Journal of Impact Engineering. 2021;**152**:103846. DOI: 10.1016/j.ijimpeng.2021.103846

[41] Munro RG. Material properties of titanium Diboride volume. Journal of Research of the National Institute of Standards and Technology. 2000;**105**:709

[42] Raju GB, Basu B, Tak NH, Cho SJ. Temperature dependent hardness and strength properties of TiB2 with TiSi2 sinter-aid. Journal of the European Ceramic Society. 2009;**29**:2119-2128. DOI: 10.1016/j.jeurceramsoc.2008.11.018

[43] Demirskyi D, Nishimura T, Sakka Y, Vasylkiv O. High-strength TiB2-TaC ceramic composites prepared using reactive spark plasma consolidation. Ceramics International. 2016;**42**:1298-1306. DOI: 10.1016/j.ceramint.2015.09.065

[44] Golla BR, Bhandari T, Mukhopadhyay A, Basu B. Titanium diboride. In: Fahrenholtz WG, Wuchina EJ, Lee WE, Zhou Y (Eds.), Ultra-High Temp. Ceram. Extrem. Environ. Appl. Hoboken, New Jersey: John Wiley & Sons, Inc.; 2014: pp. 316-360. DOI: 10.1002/9781118700853.ch13

[45] Divecha AP, Fishman SG, Karmarkar SD. Silicon carbide reinforced Aluminum—A formable composite. JOM Journal of the Minerals, Metals and Materials Society. 1981;**33**:12-17. DOI: 10.1007/BF03339487

[46] Savio SG, Madhu V, Gogia AK. Ballistic performance of alumina and zirconia-toughened alumina against 7.62 Armour piercing projectile. Defence Science Journal. 2014;**64**:464-470. DOI: 10.14429/dsj.64.6745

[47] Pathak JP, Singh JK, Mohan S. Synthesis and characterisation of aluminium-silicon-silicon carbide composite. Indian Journal of Engineering and Materials Science. 2006;**13**:238-246

[48] Oliveira MJ, Gomes AV, Pimenta AR, da S Figueiredo ABH. Alumina and low density polyethylene composite for ballistics applications. Journal of Materials Research and Technology. 2021;**14**:1791-1799. DOI: 10.1016/j.jmrt.2021.07.069

[49] Basu B, Raju GB, Suri AK. Processing and properties of monolithic TiB2 based materials. International Materials Review. 2006;**51**:352-374. DOI: 10.1179/174328006X102529

Section 3

# Refractory Ceramics

Chapter 4

# Structural, Magnetic, and Magnetodielectric Properties of Bi-Based Modified Ceramic Composites

*Rasmita Jena, Kouru Chandrakanta and Anil Kumar Singh*


## Abstract

In this chapter, we introduce a promising composite material, which can be used as a potential candidate in the field of charge storage, sensors, and spintronic devices. The structural, magnetic, and magnetodielectric properties of the pure cum composite samples are investigated. The Rietveld refinement of the X-ray data confirmed the presence of a single ($A2_1am$) and mixed phases ($A2_1am$ + $R\text{-}3c$ + $Pbam$) in the pure and composite sample, correspondingly. The SEM microstructure suggests the contrasting nature of the homogeneous and heterogeneous distribution of grains in the corresponding pure and composite sample. The magnetic properties of the composite sample increase due to the enhanced exchange interaction between the different magnetic ions. The frequency-dependent dielectric subjected to a constant magnetic field indicates the signature of magnetodielectric (MD) coupling for both the samples. The field variation of the MD loop shows the symmetric hysteresis loop in the composite due to the addition of magnetostrictive $La_{0.67}Sr_{0.33}MnO_3$ and the non-collinear antiferromagnetic $Bi_2Fe_4O_9$ phase. The maximum value of MD% (~0.12%) is enhanced by ~13 times in the composite than in the pure sample. Therefore, the improved MD coupling and symmetric switching of the MD loop of the composite make it a suitable candidate for low power consumption storage devices.


**Keywords:** composite, sol–gel modified method, magnetodielectric, Magnetoloss

## 1. Introduction

Ceramic materials are in demand due to their growing use in energy harvesting technologies, including batteries, capacitors, and storage devices [1]. There are two types of ceramic materials, i.e., traditional and advanced. The advanced ceramic material plays a significant role in the sensor and storage devices due to its high piezoelectric and resistive properties. These materials include oxides, nitrates, and carbides [2]. In these ceramics, the unique ferroelectric, electrochemical, pyroelectric, and piezoelectric properties are often useful for multiferroic research. The multiferroic materials with the simultaneous occurrence of various ferroic orders such as ferro (electric/magnetic), antiferromagnetic (AFM), ferrotoridic, and ferroelastic play a significant role in developing new technological and device applications [3]. In the recent past, researchers are focused on device miniaturization, which satisfies the vast need to integrate the electric and magnetic properties in a material. There are a variety of coupling mechanisms between the electric and magnetic orders, but magnetoelectric (ME) coupling is crucial for future micro/nanoscale electronics, low-power memory devices, and spintronic devices [4]. The ME effect was initially studied in the $Cr_2O_3$ single-phase compound. After that, many materials such as $DyMn_2O_5$, $TbMnO_3$, and $BiFeO_3$ showed the ME coupling in its single-phase [5]. However, these materials restrict their practical suitability applications due to the weak coupling among the electric and magnetic order parameters and the transition temperature below the room temperature (RT). To avoid the above difficulties in the single-phase materials, many researchers have focused their study on designing the composite materials [6]. Usually, in composite, the electric and magnetic properties are intentionally improved by adding the required electric and magnetic phases. The induced ME coupling in the composite is the product property relation between the constituent phases. The relation between the magnetic and electric phases is written as [7]:

$$ME_E = \frac{\text{electric}}{\text{mechanical}} \times \frac{\text{mechanical}}{\text{magnetic}} \text{ and } ME_H = \frac{\text{magnetic}}{\text{mechanical}} \times \frac{\text{mechanical}}{\text{electric}} \quad (1)$$

The origin of ME coupling in the composite may be strain, charge, and exchange bias mediated. It depends upon the coupling interaction at the magnetic and electric phase interface.

The alternative approach to study the coupling among the magnetic and electric ordering is the magnetodielectric (MD) effect. The existence of ME coupling can indirectly address through the MD effect. This phenomenon is defined as the magnetic field-controlled dielectric properties and reversely electric field-induced magnetic permeability [8]. Materials having MD characteristics are rich in physical content to take further research and its practical utilization. Usually, the signature of the MD effect can be realized by observing the anomaly of magnetic/dielectric transition in the dielectric/magnetic properties. The MD effect can be experienced experimentally by measuring the capacitance at the different external magnetic fields. The microscopic source of the MD effect can be originated from the extrinsic and intrinsic mechanisms. It solely depends on the origin of the dielectric properties of the material. According to G. Catalan, the MD effect can arise without having the dielectric and magnetic coupling in the sample [9]. The extrinsic mechanism responsible for the origin of the MD effect is the magnetoresistance and Maxwell-Wagner effect of the sample. Similarly, the intrinsic source of the MD effect originated from the magnetic field-induced dipolar switching mechanism. Hence, the existence of an intrinsic MD effect in a material indicates the possible signal of ME coupling. The realization of ME coupling is restricted by the symmetry requirements. The MD materials are fascinating due to their multiple microscopic origins and simplicity for device application. Recently, the MD coupling has been used to characterize the magnetic multipole orders and quantum criticality [10]. Therefore, it is necessary to investigate the MD coupling and the improvement of dielectric properties with the applied magnetic field.

The Aurivillius compound is composed of the perovskite layer $(A_{n-1}B_nO_{3n+1})^{2-}$ sandwich periodically between the $(Bi_2O_2)^{2+}$ fluorite layer. Here, n represents the number of perovskite layers present in the compound. For $n = 4$, Aurivillius compound $Bi_5Ti_3FeO_{15}$ (BTFO) is explored theoretically using the first principle calculation and experimentally [11]. The BTFO compound exhibits the orthorhombic crystallographic structure with the $A2_1am$ space group at RT. BTFO undergoes the structural transition at high temperature from ferroelectric $A2_1am$ transform to paraelectric *I4/mmm* at 730°C [12]. The BTFO has dragged the wide attention of researchers due to its high ferroelectric and piezoelectric properties above the RT. The single-phase BTFO shows the weak MD coupling at RT due to the unavailability of strong magnetic ordering. So, the artificially mixing magnetic phases in the form of the composite may provide a potential path to improve both magnetic and MD coupling in the sample. The first chosen material is $La_{0.67}Sr_{0.33}MnO_3$ (LSMO) to make the composite with the BTFO compound. Due to its exciting properties, i.e., high ferromagnetic ordering temperature ~370 K, colossal magnetoresistance, and high carrier spin polarization [13]. The second compound of interest is the $Bi_2Fe_4O_9$ (BFO), which has a unique spin frustration due to the interaction among the Fe ions. BFO ceramic shows nearly RT multiferroic behavior due to the high AFM ordering temperatures of ~260 K [14]. Therefore, the above properties of both LSMO and BFO compounds may play a pivotal role in improving the magnetic as well as MD behavior of the composite.

In this work, we have examined the physical properties of the $0.5Bi_5Ti_3FeO_{15}$-$0.2La_{0.67}Sr_{0.33}MnO_3$-$0.3Bi_2Fe_4O_9$ composite and compared it with the pure BTFO sample. The composite sample is synthesized by the sol–gel-modified technique and their dielectric, magnetic, and the source of MD effect are discussed. MD coupling in composite might be used as a potential candidate for MD device design.

## 2. Experimental

### 2.1 Synthesis of the composite samples

The ceramic composites ($0.5Bi_5Ti_3FeO_{15}$-$0.2La_{0.67}Sr_{0.33}MnO_3$-$0.3Bi_2Fe_4O_9$) were prepared by the sol–gel modified method. At first, BTFO, LSMO, and BFO samples were synthesized separately via a sol–gel auto combustion process. The chemical reagent used for the preparation of the BTFO sample were bismuth nitrate ($Bi(NO_3)_3.5H_2O$), iron nitrate ($Fe(NO_3)_3.9H_2O$), and titanium isopropoxide ($TiC_{12}H_{28}O_4$). The above chemicals were taken from Sigma-Aldrich with greater than 99.9% purity form. The deionized water was used to mix the nitrates. Bismuth nitrate and titanium isopropoxide were immiscible with the deionized water. The required drop of nitric acid was used to dissolve the chemicals and get the transparent white-colored solution. After that, add the iron nitrate to the above solution and heat it on the hot plate at 100°C. Then the desired ratio of ethylene glycol and citric acid (1:1.5) was added to the mixture. All the reagents are mixed completely and lead to the formation of the gel. The xerogel was dried on the hot plate overnight and crushed to form the homogenous powder. The resultant powder was pre-sintered in a tubular furnace at 600°C for 2 h. A similar method was used for the preparation of LSMO and BFO samples. Only the difference in the precursor materials and pre-sintered temperatures. The starting materials such as lanthanum nitrate ($La(NO_3)_3.6H_2O$), manganese nitrate ($Mn(NO_3)_2.4H_2O$), strontium nitrate ($Sr(NO_3)_2$) and bismuth nitrate, iron nitrate were used for the preparation of LSMO and BFO sample, respectively. The obtained xerogel powder was pre-sintered at 800°C for 4 h and 800°C for 2 h for the LSMO and BFO samples, respectively.

Finally, the composite ($0.5Bi_5Ti_3FeO_{15}$-$0.2La_{0.67}Sr_{0.33}MnO_3$-$0.3Bi_2Fe_4O_9$) was prepared by taking the desired ratio of as-synthesized BTFO, LSMO, and BFO powder. The proper weight percentage of 50% BTFO:20% LSMO:30% BFO (abbreviated as BLB523) was taken and mixed thoroughly with the help of agate mortar and pestle to get the homogeneous powder. The obtained powder was pressed to form the pellet and finally sintered at 900°C for 4 h.

### 2.2 Characterization

The phase identification of the pure cum composite samples was analyzed by using the X-ray diffraction (XRD) system with an ULTIMA-IV diffractometer of Cu source of radiation. The diffraction data were taken in the range of 20–60° with a slow scan rate of 3 degrees per minute. The surface morphology and elemental analysis were characterized through the scanning electron microscope (SEM) coupled with the energy dispersive X-ray spectrometer (EDAX). The size of the grains was estimated from the Image J Software. The samples magnetic properties were studied at room temperature (RT) using the vibrating sample magnetometer (VSM) with a maximum magnetic field of 15 kOe. For the electrical measurement samples were painted with high-quality silver paste to form an electrode. The RT frequency variation of dielectric permittivity at a constant magnetic field (0 to 1.3 T with a difference of 0.2 T) was studied through the impedance analyzer (Waynee Kerr 6500B model). The magnetic field variation of MD and magnetoloss was recorded by an impedance analyzer, which is assembled with the closed cycle refrigerator (CCR) system, KEPECO power supply, and the electromagnet (GMW 5034).

## 3. Results and discussion

### 3.1 Structural characterization by XRD

**Figure 1(a)** and **(b)** illustrate the XRD patterns of the pure BTFO and composite (BLB523) samples. The measured XRD data is examined through the Rietveld refinement procedure using the Fullprof software. The refinement result provides

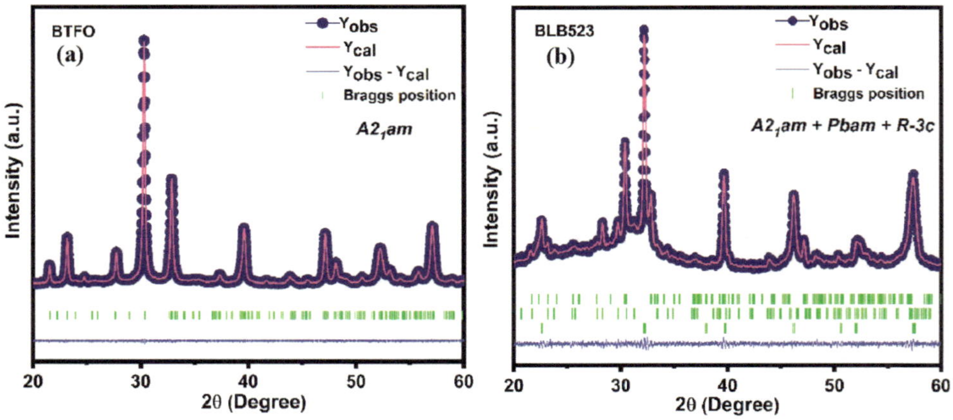


**Figure 1.**
*Rietveld refinement of the XRD patterns of (a) pure (BTFO) and (b) composite (BLB523) sample.*

| BTFO (pure) | | BLB523 (composite) | | |
|---|---|---|---|---|
| **Parameters** | | **BTFO phase** | **LSMO phase** | **BFO phase** |
| Crystal system | Orthorhombic | Orthorhombic | Rhombohedral | Orthorhombic |
| Space group | $A2_1am$ | $A2_1am$ | *R-3c* | *Pbam* |
| Lattice parameters (Å) | $a$ = 5.4555(11) | $a$ = 5.4537(9) | $a$ = 5.5618(8) | $a$ = 7.6813(9) |
| | $b$ = 5.4467(11) | $b$ = 5.4491(5) | $b$ = 5.5618(8) | $b$ = 8.5883(6) |
| | $c$ = 41.2130(7) | $c$ = 41.0190(4) | $c$ = 13.600(6) | $c$ = 5.8567(11) |
| Cell volume (Å$^3$) | 1224.63(10) | 1218.995(6) | 364.33(7) | 386.36(9) |
| Phase fraction | 100 | 50.79 | 20.63 | 28.58 |
| $R_p$ (%) | 2.20 | | 10.4 | |
| $R_{wp}$ (%) | 2.25 | | 9.65 | |
| $\chi^2$ | 1.84 | | 3.07 | |
| Crystallite size (nm) | 41.76 | | 51.35 | |
| Strain | $10.2 \times 10^{-4}$ | | $18.2 \times 10^{-4}$ | |

**Table 1.**
*Summary of the refined lattice parameters, crystallite size and strain of the BTFO and BLB523 composite.*

information related to the samples pure phase formation and structural parameters. For the pure BTFO sample, the single-phase refinement method is performed by considering the orthorhombic crystal structure ($A2_1am$ space group). In contrast, the tri-phase method is incorporated in order to refine the whole XRD pattern of the composite. The orthorhombic ($A2_1am$) BTFO, rhombohedral (*R-3c*) LSMO, and orthorhombic (*Pbam*) BFO symmetry is provided as input sources during the composite refinement. Initially, the instrumental zero correction factor is refined in the refinement process, followed by the scale factor, cell parameters ($a$, $b$, and $c$), FWHM parameters ($u$, $v$, and $w$), background points, and atomic positions. The peak shape parameters and background points are fitted by providing the Pseudo-Voigt function and linear interpolation between the background points. After several refinement cycles, the theoretically simulated pattern is well matched with the experimental data points. Additionally, the lower value of $\chi^2$, $R_p$, and $R_{wp}$ also confirm the good fitting of the theoretical model with the observed data points. The extracted structural parameters and phase fractions are presented in **Table 1**. It is observed that the lattice parameters of the composite phase show a slight deviation from that of the pure phase. It signifies the generation of lattice strain at the BTFO, LSMO, and BFO domain interfaces. The other composites also report a similar kind of variation [15]. The proposed phase fraction of the composite is well consistent with the refinement data. Hence, the existence of BTFO, LSMO, and BFO phases in the composite assure the formation of the BLB523 composite.

The mean lattice strain and crystallite size of the pure cum composite samples have been extracted from the Williamson-Hall (W-H) plot method. The mathematical expression of the W-H method is

$$\beta\cos\theta = \frac{K\lambda}{D} + 4\varepsilon\sin\theta \quad (2)$$

here θ denotes the Brrag's angle, ε is the lattice strain, λ is the incident wavelength of X-ray, ***K*** is the shape parameter (0.89 for spherical shape), *D* is the crystallite size in average, β denotes the full-width half maxima of the diffraction peak [16]. The estimated crystallite size and strain values are listed in **Table 1**. It is observed that the composite exhibits more strain than the pure sample.

### 3.2 SEM study

The surface micrograph of sintered pure cum composite samples are shown in **Figure 2(a)** and **(b)**. These images reveal the existence of grains with different sizes and orientations in the samples. The pure sample consists of well-structured plate-like grains of various sizes distributed uniformly on the sample surface. At the same time, the composite consists of a mixture of three phases (BTFO, LSMO, and BFO). As LSMO grains exist in the nanometer range, it is difficult to identify from the SEM image. The average grain size of the pure and composite sample (mixed grains) is found to be 1.44 and 0.54 μm, respectively, which is estimated from the Image J software. The EDAX analysis is incorporated to analyze the sample's elemental composition, as shown in **Figure 2(c)** and **(d)**. The EDAX spectrum indicates the constituent elemental peaks of the Bi, Ti, Fe, O and Bi, Ti, Fe, La, Sr., Mn, and O for pure and composite samples. This analysis confirms the presence of the required element in the samples.

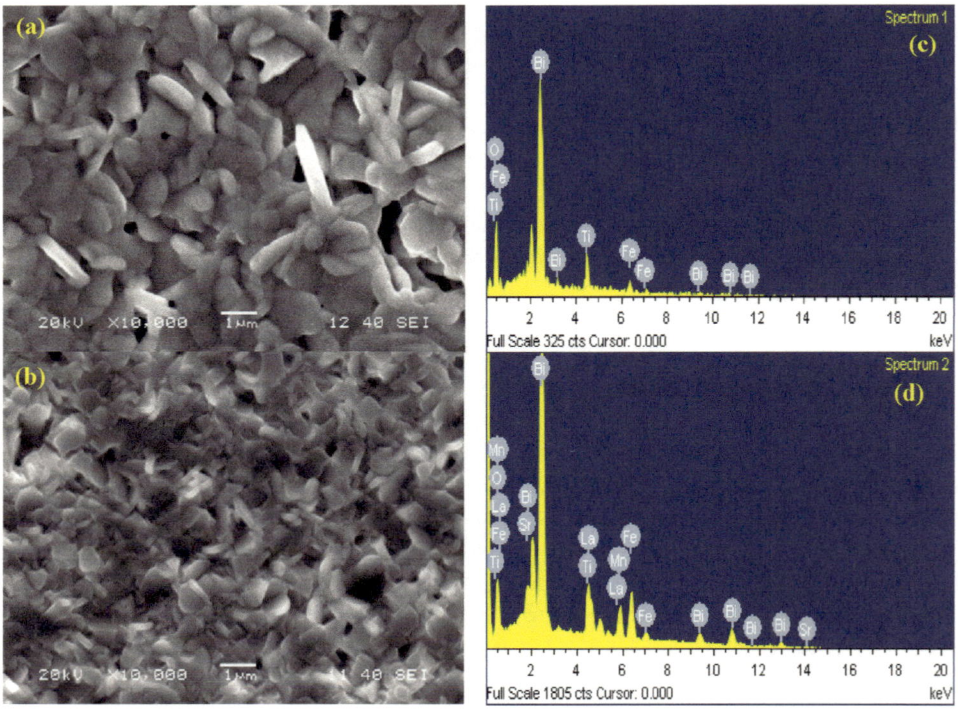


**Figure 2.**
*(a) and (b) Backscattered SEM images of sintered BTFO and BLB523 samples, respectively. (c) and (d) The EDAX spectra of the corresponding pure and composite sample.*

### 3.3 Magnetic study

The magnetization (M) versus magnetic field (H) loop for pure BTFO and composite BLB523 samples are recorded at RT. **Figure 3(a)** and **(b)** display the linear and slightly non-linear M-H loop of pure cum composite samples. This linear behavior of the M-H loop indicates the paramagnetic (PM) behavior of the pure sample. With the addition of magnetic LSMO and BFO phase, the M-H loop slightly changes to the non-linear behavior with a small opening in the field range of ±1 kOe, as shown lower inset of **Figure 3(b)**. This behavior signifies the weak ferromagnetic (FM) nature of composite. In composite BLB523, saturation magnetization is not achieved even in a high magnetic field of 15 kOe. It indicates the presence of canted spins in the composite. Since the composite sample consists of the three magnetic phases, i.e., PM BTFO, FM LSMO, and AFM BFO phases [17, 18]. The overall magnetic behavior of the composite shows the FM behavior. As the pure sample exhibit the paramagnetic behavior, with the addition of magnetic Mn ion (LSMO) and Fe ion (BFO) to the BTFO phase. The magnetic moment of the composite sample is enhanced. The different magnetic parameters such as maximum magnetization ($M_{max}$), coercive field ($H_c$), and remanent magnetization ($M_r$) are extracted from the fitting of M-H loop [19] and listed in **Table 2**. It is observed that the magnetic parameters ($M_{max}$, $H_c$, and $M_r$) increase by adding an extra magnetic phase to the pure sample. Since the maximum magnetization is an intrinsic property of the sample, which depends upon the spin configuration and spin–spin interaction between the magnetic ions. Whereas, the coercive field and remanent magnetization is affected by extrinsic factors such as particle morphology, domain structure, disorder, defects, etc. [20].

In the present discussion, the coercivity value of the BLB523 composite is increased to 145 Oe (nearly three times) than the pure sample. It could be due to the addition of manganite (LSMO) and ferrite (BFO) phase leading to the hindrance of domain interaction and resulting in the pinning effect in the composite. A similar trend of coercivity is also observed in many composites [21]. The enhanced magnetization in the composite sample is addressed due to adding extra magnetic ions, i.e., Mn and Fe ions of LSMO and BFO phase, respectively. Additionally, the enhanced

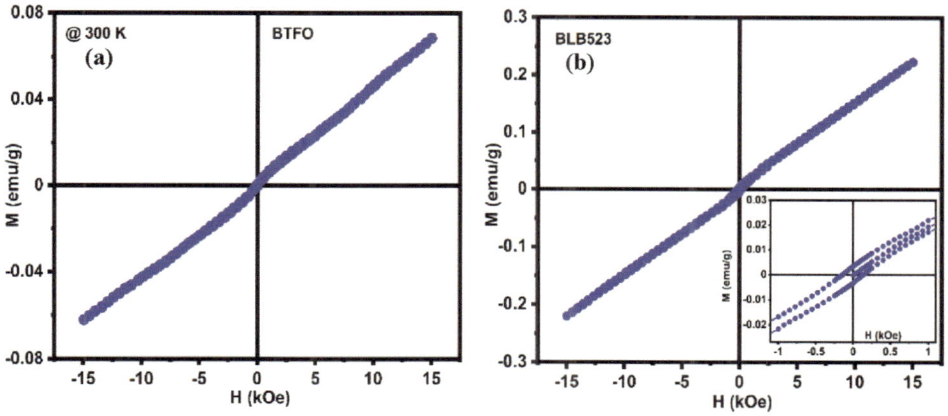


**Figure 3.**
*Magnetic hysteresis loops (M-H) of (a) BTFO and (b) BLB523 at RT.*

| Sample | $M_{max}$ (emu/g) | $M_r$ (emu/g) | $H_c$ (Oe) | $M_r/M_s$ |
|---|---|---|---|---|
| BTFO | 0.07 ± 0.02 | $3.44 \times 10^{-4}$ $\pm 6.12 \times 10^{-6}$ | 50 ± 8 | 0.005 |
| BLB523 | 0.23 ± 0.06 | $35.74 \times 10^{-4}$ $\pm 8.37 \times 10^{-6}$ | 145 ± 21 | 0.016 |

**Table 2.**
*Magnetic parameters are estimated from the M-H hysteresis loop.*

magnetization is supplemented by inherent magnetization arising due to *d-d*, *f-d*, and *f-f* exchange interaction between the Fe-Fe, Fe-Mn, and Mn-Mn ions [22]. The squareness ratio is estimated from the ratio of remanent and saturation magnetization of the samples and is given in **Table 2**. Usually, the ratio of ($M_r/M_S$) ≥ 0.5 indicates the single-domain structure, whereas ($M_r/M_S$) < 0.5 signifies the multi-domain nature of the sample [20]. In the present composite, the squareness ratio lower than 0.5 exhibits the multi-domain nature of the sample.

### 3.4 Dielectric study

The frequency variation of dielectric permittivity ($\varepsilon'$) at different external magnetic fields displays indirect access to define the magnetoelectric coupling of the material. There are the various techniques to observe the magnetodielectric effect of the sample, such as (i) the relative change of capacitance/impedance with the application of an external magnetic field and (ii) the appearance of magnetic/dielectric transition temperature in the temperature variation of dielectric/magnetic studies. In the present pure cum composite system, we have preferred the first technique to analyze the MD effect. The RT frequency dependence dielectric permittivity at a fixed magnetic field (0 T to 1.3 T with a difference of 0.2 T) is illustrated in **Figure 4(a)** and **(b)**. Both the sample shows the appreciable change in capacitance/dielectric under the application of field up to 1.3 T. A close view of the change in dielectric permittivity is plotted in the upper inset of the pure (BTFO) and composite (BLB523) samples. The reduction of $\varepsilon'$ with the change in a magnetic field signifies the presence of negative MD coupling in

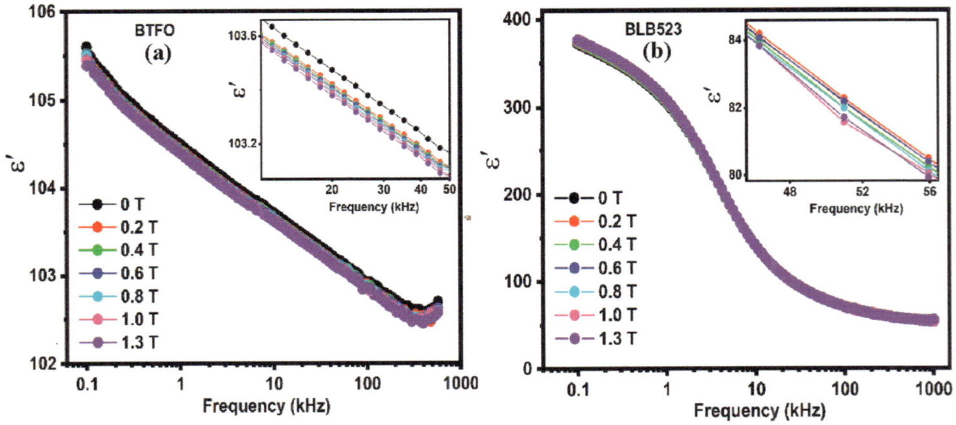


**Figure 4.**
*The frequency dependence of dielectric permittivity at the different magnetic fields of 0 T to 1.3 T (a) pure BTFO and (b) composite BLB523.*

the sample. The appearance of the positive or negative sign of the MD effect depends on the neighboring spin pair correlation and the coupling constant [23]. The dielectric constant of both the samples decreases with the increase of frequency, suggesting the usual dielectric characteristics of the sample. The moderate change of dielectric constant under the different magnetic fields in the low-frequency regime is attributed to interfacial polarization i.e., Maxwell-Wagner polarization, space charge polarization, magnetoresistance, etc. [24]. Whereas, in the high-frequency region, the dielectric value decreases due to the intrinsic dipolar contribution and suppression of extrinsic effects. The maximum strength of the $\varepsilon'$ is found to be ~105 and 375 at 100 kHz for BTFO and BLB523 samples, respectively. It is observed that the strength of the $\varepsilon'$ increases nearly three times in the composite sample. The strength of $\varepsilon'$ is reduced towards the high-frequency side by the suppression of extrinsic effects. Hence, the frequency plays a vital role in observing the sample's dielectric properties even in the presence of the magnetic field. To suppress the extrinsic contribution towards the dielectric and magnetodielectric properties of the sample. It is pivotal to study in the high-frequency region of ≥50 kHz. The aforementioned discussion gives a signature of the existence of MD coupling in the sample. The upcoming section has demonstrated a clear view of the field variation MD effect and its possible source of origin.

### 3.5 Magnetic field-dependent MD analysis

To realize the influence of magnetic field on dielectric properties of the pure and composite samples, field variation of MD measurement is recorded at RT. The MD effect can be extracted experimentally by recording the field variation dielectric data. While for the magnetoloss (ML) effect, dielectric loss data is taken as a function of a magnetic field. Both the MD and ML percentage is estimated using the following mathematical expression [19]:

$$\mathbf{MD}(\%)=\left[\frac{\varepsilon'(\boldsymbol{H},\boldsymbol{T})}{\varepsilon'(\mathbf{0},\boldsymbol{T})}-\mathbf{1}\right]\times\mathbf{100} \tag{3}$$

$$\mathbf{ML}(\%)=\left[\frac{\tan\delta(\boldsymbol{H},\boldsymbol{T})}{\tan\delta(\mathbf{0},\boldsymbol{T})}-\mathbf{1}\right]\times\mathbf{100} \tag{4}$$

here $\varepsilon'(H,T)$, $\varepsilon'(0,T)$, $\tan\delta(H,T)$, and $\tan\delta(0,T)$ are dielectric permittivity and loss with magnetic field and zero magnetic field, respectively. **Figure 5(a)–(d)** illustrates the magnetic field variation of MD% and ML% of pure cum composite samples at a constant frequency of 50 kHz. The chosen high frequency plays a vital role in exploring the extrinsic and intrinsic origin of the MD effect. The extrinsic origin arises due to the Maxwell-Wagner polarization and magnetoresistance of the sample. At the same time, the intrinsic contribution arises due to the dipolar polarization. To exclude the unwanted extrinsic contribution from the observed MD effect, the observed MD data is taken at a high-frequency region (50 kHz). The pure and composite samples show a completely different MD loop with the maximum field sweep of ±13 kOe. The BTFO pure sample shows the linear nature of the MD effect with field variation. It is because of the dominating contribution attributed by the space charge polarization. Interestingly, with the addition of the LSMO and BFO

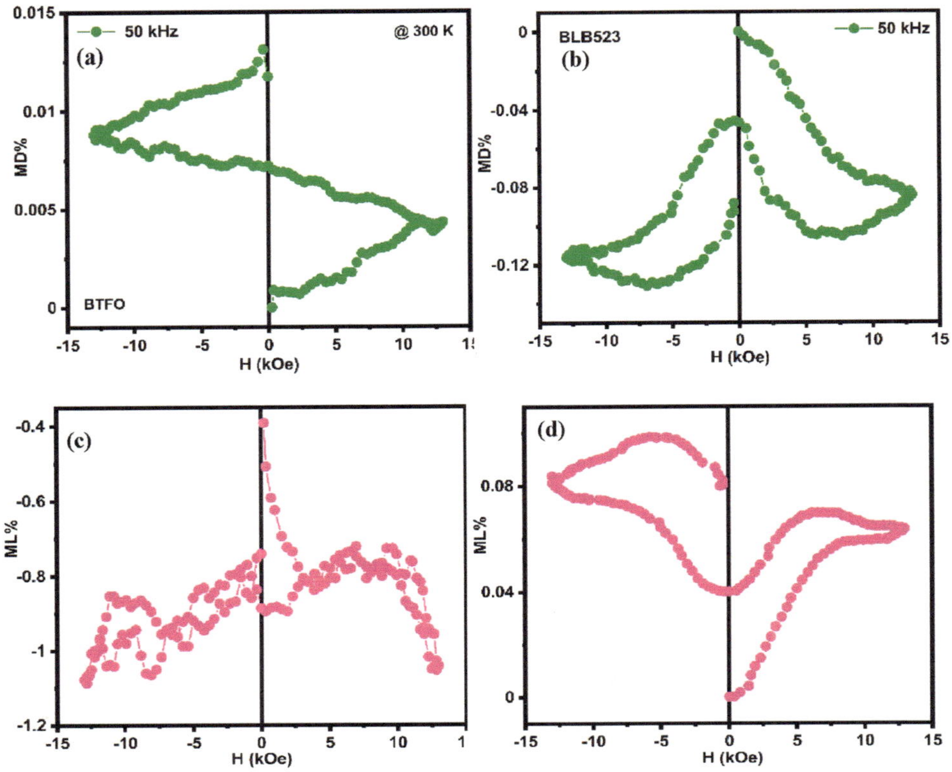


**Figure 5.**
*(a) and (b) Magnetic field variation of magnetodielectric (MD%) at a frequency of 50 kHz. (c) and (d) Magnetoloss (ML%) vs. H at 300 K for pure and composite samples.*

magnetic phase, the behavior of the MD hysteresis loop is improved. The composite sample exhibits the symmetric hysteresis loop in both positive and negative magnetic field sweeps. This MD loop is termed the inverted butterfly loop. The maximum strength of the MD% for the pure sample is ~0.01% at 13 kOe of field. With the addition of LSMO and BFO samples, the MD% of the composite sample increases to ~0.12%, which is nearly 12 times more than the pure sample. In composite, the change of MD% slope is significant around the ±8 kOe field. After that, the change of MD% tends to saturate towards the higher field of ±13 kOe. This may be due to the addition of magnetostrictive LSMO phase arising from the Mn ions spin reorientation, which makes a good mechanical coupling with the different phases. As a result, MD coupling increases up to a certain magnetic field after that, the value of magnetostriction becomes saturated with a further increase of the magnetic field. A similar kind of observation is also reported in another magnetostrictive compound [25]. According to the G. Catalan formalism, it is crucial to consider the ML effect for demonstrating the origin of the MD effect [9]. The composite sample shows the contrasting nature of ML% to that of MD%. The maximum strength of ML% (~1.08%) at 13 kOe for the pure sample is decreased to ~0.08% in the composite sample. It indicates a slight decrease in loss is observed in the composite sample. The microscopic source of the MD effect can be explained in this way: (i) the magnetostrictive LSMO phase generates the strain at the interface of the ferroelectric phase. This strain is transferred to the ferroelectric phase, which results in the capacitance of the sample [26]. (ii) the addition of the AFM BFO phase can enhance the magnetic moment due to the presence of canted spins. The non-collinear alignment of magnetic spin results in modifying the electric polarization or capacitance of the sample via inverse Dzyaloshinskii-Moriya interaction (IDM) [27]. This IDM interaction is the fundamental source behind the MD effect in the non-collinear spin structure. However, the exact source of MD coupling in the composite is still to be observed via different experimental and theoretical investigations. Hence, the symmetric switching MD hysteresis loop with good coupling strength (~0.12%) in the composite sample might be used in the charge storage devices.

## 4. Conclusion

In summary, the pure cum composite ($0.5Bi_5Ti_3FeO_{15}$-$0.2La_{0.67}Sr_{0.33}MnO_3$-$0.3Bi_2Fe_4O_9$) is successfully prepared by using the sol–gel and its modified technique. The XRD analysis confirmed the existence of single (*A2₁am*) and composite (*A2₁am*, *R-3c*, and *Pbam*) phases in the pure and composite samples, respectively. The SEM analysis has confirmed the presence of homogeneous and heterogeneous microstructure of pure and composite samples. The presence of constituent elements of different phases has been detected from the EDAX spectrum. The average grain size of the pure and mixed grains of the composite sample is found to be 1.44 and 0.54 μm, respectively. The addition of magnetic LSMO and BFO phases enhances the overall magnetic properties of the composite sample. The magnetic parameters $M_s$ and $H_c$ in the composite are enhanced by nearly three times than the parent BTFO. This is attributed to the inherent *d-d*, *f-d*, and *f-f* exchange interaction between the Fe-Fe, Fe-Mn, and Mn-Mn ions. The frequency dependence of the dielectric constant at a fixed magnetic field demonstrates a signature of the MD effect in the pure and composite samples. This observation encourages studying the MD effect in the pure cum composite sample. The field-induced maximum strength of the MD effect is about ~0.12% at 50 kHz observed in the composite sample. This MD effect can be the combined effect of strain magnetostrictive LSMO and the inverse Dzyaloshinskii-Moriya interaction generated by the non-collinear Fe ions in the AFM BFO phase. Hence, the present composite establishes a relation between the electric and magnetic order of the constituent phases and further study can explore possibilities in MD device applications.

## Acknowledgements

All authors would like to acknowledge the Board of Research in Nuclear Science (Project No: 2012/37P/40/BRNS/2145), Mumbai, the UGC-DAE CSR Mumbai (Project No: CRS/2021-2022/03/585), and the Department of Science and Technology (Project No: SR/FTP/PS-187/2011), New Delhi for funding.

## Conflict of interest

The authors declare no conflict of interest.

## Author details

Rasmita Jena, Kouru Chandrakanta and Anil Kumar Singh*
Department of Physics and Astronomy, National Institute of Technology Rourkela, Odisha, India

*Address all correspondence to: singhanil@nitrkl.ac.in

## References

[1] Cramer CL, Ionescu E, Zajac MG, Nelson AT, Katoh Y, Haslam JJ, et al. Additive manufacturing of ceramic materials for energy applications: Road map and opportunities. Journal of the European Ceramic Society. 2022;**42**:3049-3088. DOI: 10.1016/j.jeurceramsoc.2022.01.058

[2] Chalia S, Bharti MK, Thakur P, Thakur A, Sridhara SN. An overview of ceramic materials and their composites in porous media burner applications. Ceramics International. 2021;**47**:10426-10441. DOI: 10.1016/j.ceramint.2020.12.202

[3] Scott JF. Room-temperature multiferroic magnetoelectrics. NPG Asia Materials. 2013;**5**:1-11. DOI: 10.1038/am.2013.58

[4] Spaldin NA, Ramesh R. Advances in magnetoelectric multiferroics. Nature Materials. 2019;**18**:203-212. DOI: 10.1038/s41563-018-0275-2

[5] Ortega N, Ashok K, Scott JF, Katiyar RS. Multifunctional magnetoelectric materials for device applications. Journal of Physics: Condensed Matter. 2015;**27**:1-23. DOI: 10.1088/0953-8984/27/50/504002

[6] Chen J, Bai Y, Nie C, Zhao S. Strong magnetoelectric effect of $Bi_4Ti_3O_{12}$/$Bi_5Ti_3FeO_{15}$ composite films. Journal of Alloys and Compounds. 2016;**663**:480-486. DOI: 10.1016/j.jallcom.2015.12.088

[7] Nan CW, Bichurin MI, Dong S, Viehland D, Srinivasan G. Multiferroics magnetoelectric composites: Historical perspective, status, and future directions. Journal of Applied Physics. 2008;**103**:1-35. DOI: 10.1063/1.2836410

[8] Yao X, Zhou JP, Zhang XL, Chen XM. Magnetodielctric mechanism and application of magnetoelectric composites. Journal of Magnetism and Magnetic Materials. 2022;**550**:1-11. DOI: 10.1016/j.jmmm.2022.169099

[9] Catalan G. Magnetocapacitance without magnetoelectric coupling. Applied Physics Letter. 2006;**88**:1-3. DOI: 10.1063/1.2177543

[10] Huang S, Jin H, Wan KQ, Wang HO, Su KP, Yang DX, et al. The antiferromagnetic ordering and metamagnetic transition induced magnetoelectric effect in $Dy_2Cu_2O_5$. Journal of Applied Physics. 2022;**131**:1-6. DOI: 10.1063/5.0071660

[11] Birenbaum AY, Scaramucci A, Ederer C. Magnetic order in four-layered Aurivillius phases. Physical Review B. 2017;**95**:1-10. DOI: 10.1103/PhysRevB.95.104419

[12] Jena R, Chandrakanta K, Pal P, Abdullah MF, Kaushik SD, Singh AK. Dielectric relaxation and conduction mechanism in Aurivillius ceramic $Bi_5Ti_3FeO_{15}$. International Journal of Minerals, metallurgy, and Materials. 2021;**28**:1063-1071. DOI: 10.1007/s12613-020-2091-3

[13] Majumdar S, Dijken SV. Pulsed laser deposition of $La_{1-x}Sr_xMnO_3$: Thin film properties and spintronic applications. Journal of Physics D: Applied Physics. 2014;**47**:1-15. DOI: 10.1088/0022-3727/47/3/034010

[14] Pooladi M, Sharifi I, Behzadipour M. A review of the structure, magnetic, and electric properties of the bismuth ferrite ($Bi_2Fe_4O_9$). Ceramics International. 2020;**46**:18453-18463. DOI: 10.1016/j.ceramint.2020.04.241

[15] Pandey R, Pradhan LK, Kumar S, Supriya S, Singh RK, Kar M.

Correlation between lattice strain and physical (magnetic, dielectric, and magnetodielectric) properties of perovskite-spinel ($Bi_{0.85}La_{0.15}FeO_3$) (1-x)- ($NiFe_2O_4$)(x) composites. Journal of Applied Physics. 2019;**125**:1-11. DOI: 10.1063/1.5063775

[16] Verma KC, Singh M, Kotnala RK, Goyal N. Magnetic field control of polarization/capacitance/voltage/resistance through lattice strain in $BaTiO_3$-$CoFe_2O_4$ multiferroic nanocomposite. Journal of Magnetism and Magnetic Materials. 2019;**469**:483-493. DOI: 10.1016/j.jmmm.2018.09.020

[17] Ahmed MA, Bishay ST, Salem-Gaballah SM. Structural characterization and magnetic properties of smart CuCd ferrite/LaSrCo manganite nanocomposites. Journal of Magnetism and Magnetic Materials. 2013;**334**:96-101. DOI: 10.1016/j.jmmm.2013.01.022

[18] Wang Y, Xu G, Yang L, Ren Z, Wei X, Weng W, et al. Low-temperature synthesis of $Bi_2Fe_4O_9$ nanoparticles via a hydrothermal method. Ceramic International. 2009;**35**:51-53. DOI: 10.1016/j.ceramint.2007.09.114

[19] Jena R, Chandrakanta K, Abdullah MF, Pal P, Kaushik SD, Singh AK. Structural, magnetic, and magnetodielectric correlations in multiferroic $Bi_5Ti_3FeO_{15}$. Journal of Material Science: Materials in Electronics. 2021;**32**:21379-21394. DOI: 10.1007/s10854-021-06641-8

[20] Sharma S, Sharma H, Thakur S, Shah J, Kotnala RK, Negi NS. Structural, magnetic, magneto-dielectric and magneto-electric properties of (1-x) $Ba_{0.85}Ca_{0.15}Ti_{0.90}Zr_{0.10}O_3$-(x) $CoFe_2O_4$ ;lead-free multiferroic composites sintered at higher temperature. Journal of Magnetism and Magnetic Materials. 2021;**538**:1-9. DOI: 10.1016/j.jmmm.2021.168243

[21] Bharadwaj S, Tirupathi A, Kumar NP, Pola S, Lakshmi YK. Study of magnetic and magnetoresistance behaviour of $La_{0.67}Sr_{0.33}MnO_3$-$CoFe_2O_4$ composites. Journal of Magnetism and Magnetic Materials. 2020;**513**:1-9. DOI: 10.1016/j.jmmm.2020.167058

[22] Mohapatra SR, Vishwakarma PN, Kaushik SD, Singh AK. Effect of holmium substitution on the magnetic and magnetodielectric properties of multiferroic $Bi_2Fe_4O_9$. Journal of Applied Physics. 2017;**122**:1-9. DOI: 10.1063/1.4994645

[23] Chandrakanta K, Jena R, Pal P, Abdullah MF, Sahu DP, Kaushik SD, et al. Temperature-dependent magnetodielectric, magnetoimpedance, and magnetic field controlled dielectric relaxation response in $KBiFe_2O_5$. Journal of Magnetism and Magnetic Materials. 2022;**549**:1-8. DOI: 10.1016/j.jmmm.2022.169047

[24] Wu H, Xu R, Zhou C, Xing S, Zeng Z, Ao H, et al. Effect of core size on the magnetoelectric properties of $Cu_{0.8}Co_{0.2}Fe_2O_4$@$Ba_{0.8}Sr_{0.2}TiO_3$ ceramics. Journal of Physics and Chemistry of Solids. 2022;**160**:1-10. DOI: 10.1016/j.jpcs.2021.110314

[25] Pal M, Srinivas A, Asthana S. Enhanced magneto-electric properties and magnetodielectric effect in lead-free (1-x)$0.94Na_{0.5}Bi_{0.5}TiO_3$-$0.06BaTiO_3$-x$CoFe_2O_4$ particulate composites. Journal of Alloys and Compounds. 2022;**900**:1-13. DOI: 10.1016/j.jallcom.2021.163487

[26] Rather MD, Samad R, Hassan N, Want B. Magnetodielectric effect in rare-earth doped $BaTiO_3$-$CoFe_2O_4$ multiferroic composites. Journal

of Alloys and Compounds. 2019;**794**:402-416. DOI: 10.1016/j.jallcom.2019.04.244

[27] Kimura T. Spiral magnets as magnetoelectrics. The Annual Review of Materials Research. 2007;**37**:387-413. DOI: 10.1146/annurev.matsci.37.052506.084259

Section 4

# Smart Ceramic Materials

Chapter 5

# Design and Development of Zirconia-Alumina Bioceramics Obtained at Low Temperature through Eco-Friendly Technology

*Rut Benavente, Maria Dolores Salvador and Amparo Borrell*

## Abstract

Ceramics are increasingly used as structural materials with biomedical applications due to their mechanical properties, biocompatibility, esthetic characteristics and durability. Specifically, zirconia-based compounds are commonly used to develop metal-free restorations and dental implants. The consolidation of ceramics is usually carried out through powders by means of processes that require a lot of energy, as long as processing times and high temperatures (over 1400°C) are required. In the recent years, new research is being developed in this field to reduce both energy consumption and processing time of ceramic powders. One of the most promising techniques for sintering ceramics is microwave heating technology. The main objective of this chapter is to obtain highly densified zirconia-alumina compounds by microwave technology. After sintering, the materials are characterized to determine whether the final properties meet the mechanical requirements for their final applications as dental material. Finally, the characterization of specimens treated by low-temperature degradation (LTD) is carried out after each 20 h of LTD exposure up to 200 h. In addition, the quantification of monoclinic phase by micro-Raman spectroscopy, analysis by AFM and Nomarski optical microscopy and assessment of the roughness and mechanical properties (hardness and Young's modulus) by nanoindentation technique have been studied.

**Keywords:** zirconia-alumina, microwave sintering, bioceramics, mechanical properties, hydrothermal degradation

## 1. Introduction

In recently years, zirconia and alumina have been recognized as the most relevant ceramic materials due to their outstanding properties, such as hardness, fracture toughness, Young's modulus, chemical stability, wear and mechanical strength. Due to these excellent properties, $ZrO_2$ and $Al_2O_3$ are appropriate materials for a wide range of applications, such as the manufacture of sensors, fuel cells, thermal barriers, implants and structural engineering applications [1–3]. Recent studies claim that the incorporation of alumina into the zirconia matrix, alumina-reinforced zirconia (ATZ) materials, improves the mechanical properties (i.e., hardness, toughness and wear resistance) [4, 5], as these composites combine the unique properties of alumina and zirconia. These characteristics make $ZrO_2$ and $Al_2O_3$ promising composites for prosthetic and dental implant applications. There are a number of reports on the sintering of ATZ composites in the literature. Li et al. [4] have sintered a $ZrO_2$ (3YTZP) + 20% wt% $Al_2O_3$ composite by spark plasma sintering (SPS) a non-conventional sintering technique. The hardness and fracture toughness values achieved for these samples sintered by SPS at 1400°C were 12.5 GPa and 5.3 MPa-m$^{1/2}$, respectively. The ATZ ($ZrO_2$ (3YTZP) + 10% vol. $Al_2O_3$) composite was also studied by Meena and Karunakar [5]; in this case, the maximum hardness values for samples sintered by SPS at 1300°C were 19.8 GPa. In summary, the final properties and microstructure of the materials depend on the densification process of the material, sintering mechanisms and methods [6–9].

Substantial improvements in dental prostheses and implants have been achieved through the employment of ceramic-based materials, mainly thanks to the advent of yttria-stabilized zirconia polycrystalline (Y-TZP) as biomaterials [10–13]. The replacement of metal parts in orthopedic and dental applications with ceramics is currently on the rise. Ceramic materials provide several advantages over metals, such as biocompatibility, outstanding mechanical properties and esthetics.

However, an important characteristic of these materials is completely stabilized zirconia tetragonal (t) phase with the addition of an oxide ($Y_2O_3$, $CeO_2$,...), for example, with ~3.0 mol% of yttria for dental applications. This is important because when Y-TZP materials are subjected to humidity at 25–280°C [14–16], a sharp decline in their mechanical properties occurs over time. This phenomenon is referred to as low-temperature degradation (LTD) [17, 18]. The t-m transformation produces volume changes and defects in the material, and the mechanical and esthetic properties are affected. Thus, it is highly relevant to research the susceptibility of Y-TZP-based materials to LTD. LTD process is influenced by different factors, such as the tetragonal stabilizing dopant, the grain size or the porosity. Several of these elements are related to the process of sintering and its heating mechanisms.

In the late 1990s, the Y-TZP femoral heads used for the hip replacement failed catastrophically within the human body and these failures were attributed to the hydrothermal aging process [19, 20]. The conditions that promote LTD are found in the dental cavity; therefore, it is critically necessary to investigate the effects of LTD on Y-TZP-based materials for these uses.

Often, ceramics are full consolidated using a thermal treatment at high temperatures (>1200°C), where the temperature and dwell time are the most significant parameters since they establish the mechanical properties and microstructure of the densified material. Conventional sintering requires long-processing times and high temperatures and consequently high-energy expenditure. For this reason, non-conventional and fast sintering methods, such as microwave heating technology, are being implemented in the last decade. Microwave sintering is founded on the absorption of electromagnetic radiation, resulting in an increase in the temperature of the material [21–23]. The mechanism of microwave heating differs from the one used in conventional sintering, since the temperature gradient is, conversely, from the inside to the outside. It is referred to as volumetric heating [24]. Other investigations have also explored the influences of the processing requirements, the addition of second phases such as $Al_2O_3$ or $Nb_2O_5$, and the incorporation of $Y_2O_3$ stabilizers on the behavior of LTD for dental materials sintered by the traditional method [25, 26].

Nevertheless, no comprehensive research has been conducted on the effect of microwave sintering on 3Y-TZP/$Al_2O_3$ composites when exposed to LTD conditions.

The aim of this research was to assess the impact of microwave sintering on the LTD resistance of dental materials based on zirconia-alumina nanocomposites by comparing them with materials sintered by the traditional technique. This includes the evaluation of surface roughening, monoclinic phase transformation progression, as well as the mechanical properties as a function of the degradation time in simulated laboratory settings.

## 2. Materials and methods

A commercial zirconia-alumina powders (ATZ) from the company Tosoh (TZ-3Y20A, Tosoh, Tokyo, Japan) have been studied in this work. Specifically, it is composed of 80 wt% of 3Y-TZP and 20 wt% of $Al_2O_3$.

The specimens have been pressed uniaxially (50 MPa of pressure load) with a universal testing machine (Shimadzu AG-X Plus) to obtain circular shapes of 10 mm (diameter) and 3 mm of height. All green samples had a geometric density of ~55% in relation to the theoretical density.

The sintering of the samples was carried out using two different techniques: microwave technology (a non-conventional fast technique) and conventional oven. Microwave sintering (MW) has been performed in a single-mode circular cavity microwave oven operating in the $TE_{111}$ mode with a resonant frequency of 2.45 GHz (**Figure 1**, [27]). The microwave sintering conditions have been 1300°C of maximum temperature during 10 min of dwell time and 50°C/min of a heating rate. It is worthy of being mentioned that the zirconia-alumina composites are poorly microwave-absorbent materials at low temperatures, since its dielectric loss factor at room temperature is less than 0.01 [28]. This fact makes hybrid heating necessary, using silicon carbide as a susceptor [29, 30]. The sample temperature is controlled by an infrared radiation pyrometer (Optris CT-Laser GH5, 5 μm), which focuses on the sample surface through the tiny circular opening in the cavity wall. The emissivity and transmissivity of the final temperature material are determined prior to sintering test.

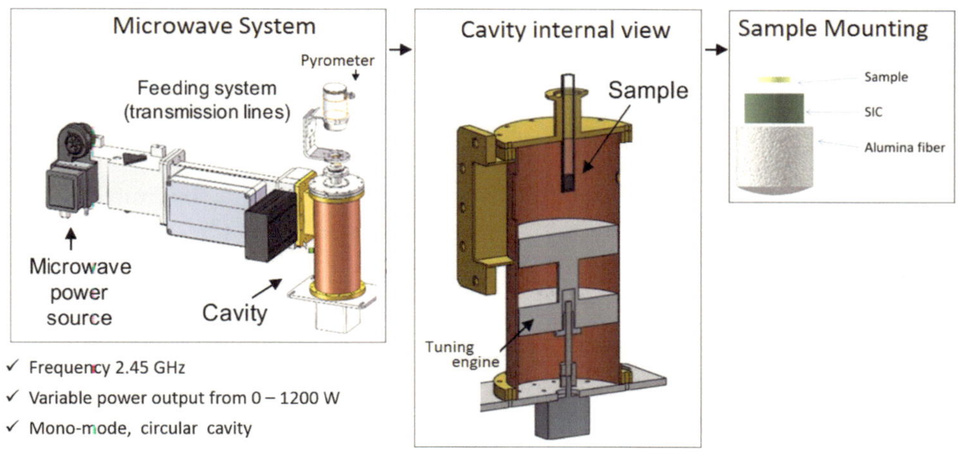


**Figure 1.**
*Details of the experimental microwave system of 1 kW at 2.45 GHz connected to a mono-mode circular cavity.*

Conventional sintering (CS) has been performed in an electric oven (Carbolite Gero, HTF 1800) for 120 min in atmospheric conditions at 1500°C and a heating rate of 10°C/min. Both the MW and CS sintering conditions have been extracted from a previous study, where their main mechanical and microstructural properties were analyzed [31]. These chosen settings are based on previous research by our group, in which the sintering conditions of zirconia-based materials were optimized [32, 33].

The study of low-temperature degradation (LTD) of $ZrO_2$-$Al_2O_3$ materials has been developed using procedures that simulate and intensify the process of hydrothermal aging. The samples are then autoclaved (MARK) with steam at 125°C and 1.6 bar [19].

The aging samples are characterized after each 20 h of exposition to LTD until they reach 200 h. This characterization consists of the following:

a. quantification of phase content by Raman spectroscopy (Horiba-MTB Xplora). The linear model suggested by Lim et al. is utilized to calculate the content of the monoclinic phases (*m*) in the samples [34]:

$$V_m = \frac{I_m^{181} + I_m^{190}}{0.33 \cdot \left(I_t^{147} + I_t^{265} + I_m^{181} + I_m^{190}\right)} \tag{1}$$

where $V_m$ is the volume fraction of monoclinic phase and $I$ is the integrated peak intensity (low peak area). The intensities of the $t$-phase are 147 and 265 cm$^{-1}$, while those of the $m$-phase are 181 and 190 cm$^{-1}$ [35].

b. assessment of the topography and roughness (Ra) of the surface characterized by atomic force microscopy (AFM Multimode, Veeco) for which samples were mirror polished with diamond paste before sintering, so that the changes in topography caused by the transformation of degradation can be effectively assessed. Two topographic photos have been obtained for every specimen: an image with a scan area of 1 μm × 1 μm for a close-up view of the grains and a second image with a scan area of 5 μm × 5 μm to provide a representative roughness value.

c. microscopic analysis of damaged surfaces by Nomarski optical microscopy. Nomarski microscopy is another method to visualize the specimen's topography in three dimensions.

d. evaluation of their mechanical characteristics (hardness and Young's modulus) according to the nanoindentation technique. The method used is a nanoindenter (G-200; Agilent Technologies). Tests were carried out with a Berkovich tip calibrated with silica standard and operated at a maximum depth of 2000 nm. The continuous stiffness measurement was used to determine the contact stiffness and calculate the hardness profiles and elastic modulus [36]. A matrix with 25 indentations was made for every material.

## 3. Results and discussion

### 3.1 Microstructure and phase transformation

The relative densities of sintered composite obtained by microwave technology at 1300°C with 10 min of dwell time (MW1300_10) and conventional oven at 1500°C

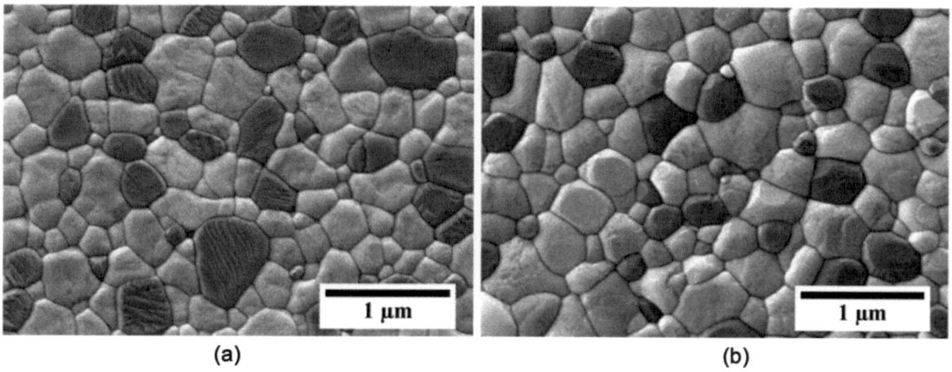


**Figure 2.**
*FE-SEM micrograph of ATZ composites under different sintering conditions: (a) MW1300_10 and (b) CS1500_120.*

with 120 min of dwell time (CS1500_120) were around 99.8 and 99.2%, respectively, and its grain sizes below 500 nm (**Figure 2**).

The relative density is calculated from the theoretical density of the ATZ sample, 4.89 g/cm$^3$. The results indicated that all specimens reached a high degree of densification. It should be noted that the samples sintered by MW at 2.45 GHz exhibited a relative higher density than samples sintered by CS, achieving 99.8% for MW1300_10. It is important to note that the sintering temperatures and holding times employed in microwave sintering are considerably lower than in a conventional process. In conclusion, while the sintering time needed to achieve relative densities above 99% with conventional sintering is 350 min, microwave technology produces denser samples in only 35 min. It should be noted that the final economic cost is considerably reduced due to the decrease in processing time and energy consumption and, consequently, the environmental impact also decreases. Therefore, microwave technology is considered to be a more environmentally friendly technique than conventional sintering.

The FE-SEM micrographs of the MW and CS densified ATZ composite are shown in **Figure 2**. The samples are very dense and have a high homogeneity, since the alumina grains (the darker ones) are uniformly dispersed in the zirconia matrix. These results are consistent with the relative density values. The darker grains correspond to alumina, while the lighter grains are zirconia. The average grain sizes of zirconia and alumina have been measured from their micrographs. The grain size increases with residence time and sintering temperature.

As for the sample densified by (MW1300_10), the average grain size of $ZrO_2$ and $Al_2O_3$ reached approximately 280 and 400 nm, respectively. In the case of the sample sintered by conventional furnace, the evolution of the average grain sizes was similar to that of MW; the higher the sintering temperature, the larger the grain size. In sample HC1500_120, the grain sizes of $ZrO_2$ and $Al_2O_3$ were 330 and 450 nm, respectively [31]. Wu et al. [37] researched the effects of $Al_2O_3$ addition in 3Y-TZP on the mechanical properties and microstructure of the composite. The increase in alumina content favored slightly the grain growth during the densification. Thus, the average grain size of $ZrO_2$ is slightly higher than that reported in the literature [38].

Raman spectra for conventional and microwave-sintered ATZ composites are shown in **Figure 3**. It is possible to check the phase transformation by Raman spectroscopy, thanks to the characteristic doublet of the *m*-phase at 181–190 cm$^{-1}$.

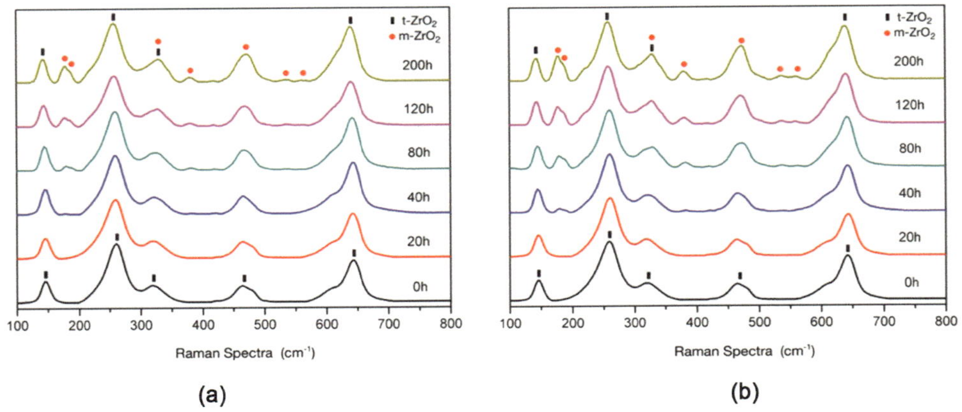


**Figure 3.**
*Raman spectrums of different LTD times of exposure for sintered ATZ material by (a) MW1300_10 and (b) CS1500_120.*

As can be seen, no monoclinic phase peaks are present in any of the samples after sintering (0 h of exposure to LTD) and also after 20 h of LTD exposure. As ATZ samples are exposed to LTD for longer, the intensity of the monoclinic phase peaks increases. After 40 h, such peaks can be recognized for sample CS1500_120, while for MW1300_10, they appear after 80 h. As the degradation time increases, the transformation becomes rather important with a distinct presence at 181–190 $cm^{-1}$.

If both sintering methods are compared, differences in the intensities of the doublet characteristic of the monoclinic phase can be observed between the spectra. The monoclinic peaks are higher in the sample CS1500_120. This fact suggests a greater vulnerability to transformation induced by LTD in conventionally sintered samples than by microwave technique [39].

The phase transformation can be quantified from the volume fraction of the m-phase, Vm, measured by Raman spectra (Eq. 1). The results of the quantification of Vm are shown in **Figure 4**.

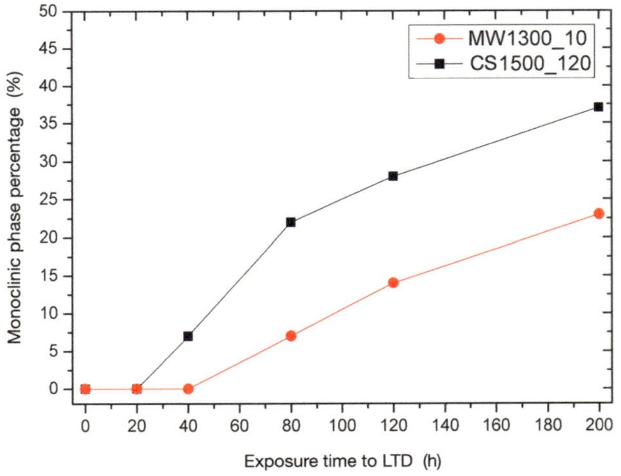


**Figure 4.**
*Volume fraction of the m-phase, Vm, a result of the LTD exposure time for ATZ material in either sintering conditions (MW1300_10 and CS1500_120).*

The lines show differences in the kinetics of the phase change. The MW1300_10 sample sintered by MW degrades more slowly compared with the conventionally densified CS1500_120 sample. After 200 h of degradation, Vm is 24% for MW1300_10 and 37% for CS1500_120. Therefore, the sample obtained by microwave is less susceptible to LTD.

After these results, it is verified that adding alumina to zirconia, forming an ATZ composite causes the zirconia to degrade more slowly compared with monolithic zirconia. Presenda et al. concluded that Y-TZP transforms approximately 90% after 100 h of testing under the same conditions as this work [17]. Ultimately, the resistance to aging is increased with the addition of alumina in the material composition.

### 3.2 Topography and surface roughness

A volume increases of about 3–4% accompany the t- to m-phase transformation of $ZrO_2$-based composites. **Figure 5** shows the AFM images of the ATZ composite sintered by CS and MW, where it is possible to analyze the surface changes induced by LTD exposure. The average surface roughness, Ra, has also been identified as a way to measure this variation.

As the exposure time increases, the topography becomes more irregular, increasing the surface roughness. After 200 hours of exposure to LTD, the surfaces of the samples tested are found to have increased in roughness, with the appearance of bulging.

In purpose of comparing the roughness of the samples sintered by different sintering methods, the Ra values at different exposure times have been determined. These values are presented in **Figure 6.**

Sample MW1300_10 has given lower Ra values than CS1500_120, showing greater variability in rugosity, particularly beyond the first 20 h. A strong increase

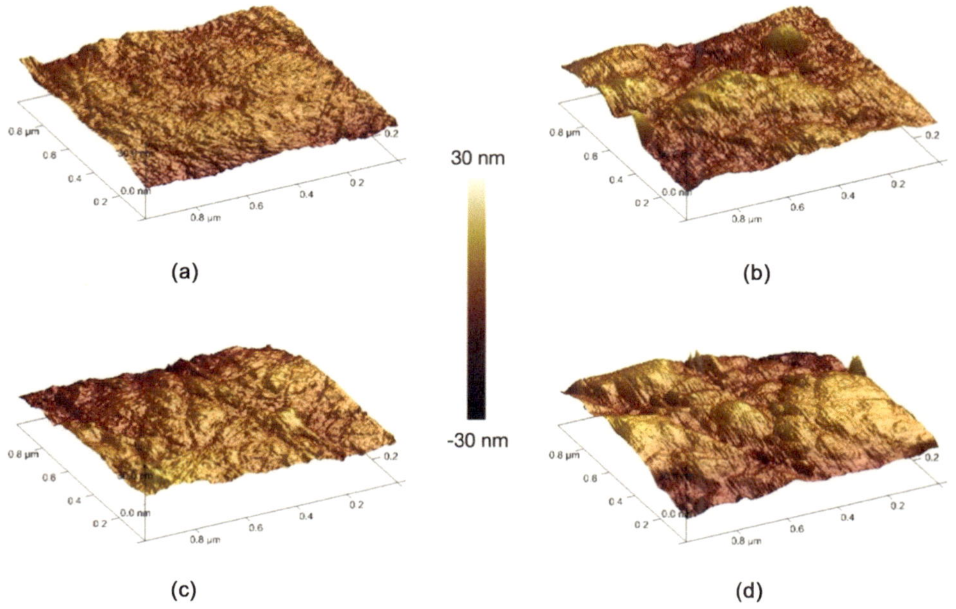


**Figure 5.**
*AFM topographic images of ATZ composite at various exposure times of LTD for MW1300_10: (a) 0 h, (b) 200 h; and CS1500_120: (c) 0 h and (d) 200 h.*

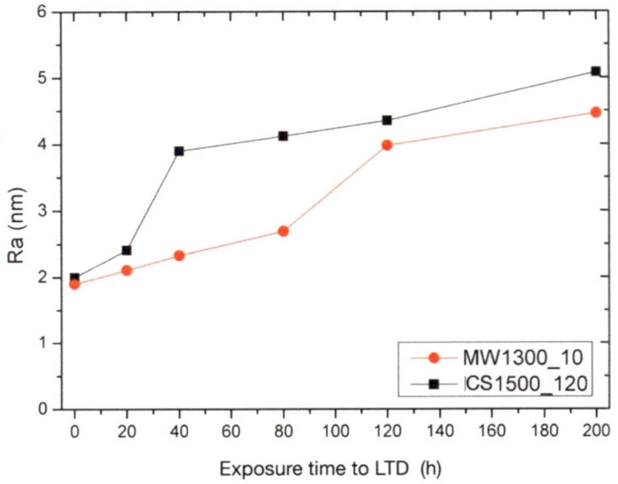


**Figure 6.**
*Average surface roughening at various LTD times for each specimen.*

is observed after the first 20 h for CS1500_120, reaching about 5.1 nm after 200 h of LTD. However, it was not until after 80 h that a sharp change in Ra was observed in sample MW1300_10. This performance suggests that the more meaningful topographical changes vary depending on the sintering method.

These results are consistent with the transformation of the $t$-phase to $m$-phase according to the Raman spectra obtained (**Figure 3**); because after 80 h, the monoclinic peak begins to be observed in the Raman spectra for the sample MW1300_10, which is when the sudden jump in the Ra values appears. The same happens for the sample CS1500_120.

Although in the Raman spectra of both samples there are no monoclinic peaks until after 20 h of LTD exposure, by observing the surface changes at the submicrometric scale with AFM, it can be seen these microstructure changes are occurring before the peaks appear. As the exposure time of the samples to LTD increases, the irregularities in the surface of the samples also increase because of the push of the grains to the surface to explain the expansion in volume.

### 3.3 Nomarski optical microscopy

Nomarski optical microscopy is another technique used to study degraded surfaces. **Figure 7** shows the micrographs of the surfaces of the samples just after sintering (at 0 h) and after 80 h of LTD time of exposure.

The effects of degradation appear as surface roughness by volume due to phase transformation. After 80 h, changes are already observed on the surface, where an increase in roughness is observed. It is true that the irregularities, as well as the phase transformation, are lower than those observed in monolithic zirconia samples, such as, for example, Y-TZP [17]. Being the yttria in both materials the dopant of zirconia, the addition of alumina in the composition is what makes the aging of zirconia difficult.

### 3.4 Effect on mechanical properties

After exposure to LTD, the structural quality of the material is evaluated, analyzing the hardness and Young's modulus, depending on the aging time. The tests were performed on the polished surfaces that were exposed to LTD. The values are shown in **Figure 8**.

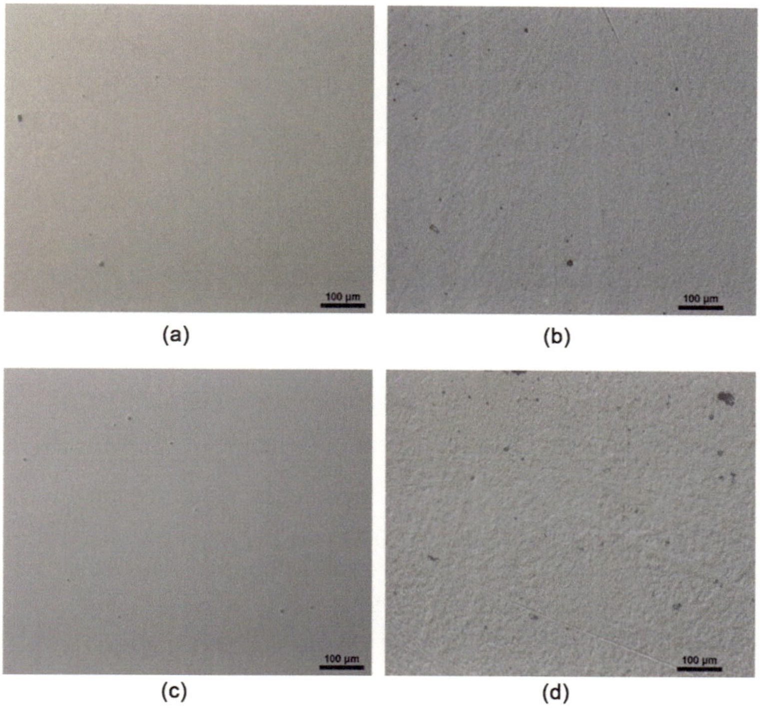


**Figure 7.**
*Nomarski microscopy images of ATZ composite in various exposure times LTD for MW1300_10: (a) 0 h, (b) 80 h; and CS1500_120: (c) 0 h and (d) 80 h.*

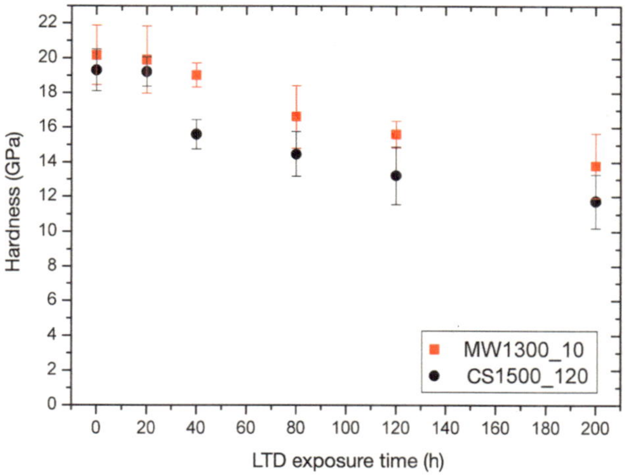


**Figure 8.**
*Hardness values for samples as a function of the LTD time of exposure.*

As can be seen, the hardness is affected differently by the sintering method used, with the sintered sample offering more resistance to wear by MW, which corroborates the results obtained in the other techniques. After sintering, before exposure to LTD, the samples show a hardness value of 20.3 and 19.5 GPa for MW1300_10 and CS1500_120, respectively. At 40 h, the hardness for CS1500_120 drops sharply to 15.7 GPa. However, the sample MW1300_10 decreases its hardness more slowly,

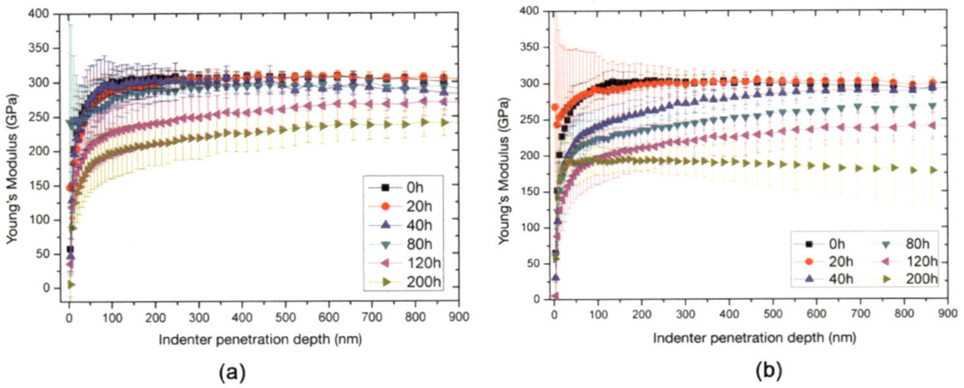


**Figure 9.**
*Young's module values for various LTD exposure of times for sintered ATZ composite by (a) MW1300_10 and (b) CS1500_120.*

following a less pronounced trend. After 200 h, the hardness values obtained are 13.8 and 11.9 GPa for MW1300_10 and CS1500_120, respectively. These results show that MW1300_10 resists better the exposure to hydrothermal degradation.

**Figure 9** shows Young's modulus values for the composite densified by both sintering methods.

Like hardness, Young's modulus (E) is affected after exposure to LTD. The E values are modified in both samples, although for MW1300_10 to a lesser extent than CS1500_120. The first significant change in the E-value appears at 40 h for CS1500_120, while in sample MW1300_10 such change is not observed until after 80 h, corroborating the results obtained previously. After 200 h, the average values of Young's modulus are 237 and 170 GPa for MW1300_10 and CS1500_120, respectively.

So far, a number of studies have concentrated on the understanding of the LTD process and its effect on 3Y-TZP dental materials sintered *via* conventional methods [40]. The investigation has also explored the influence of processing conditions, the incorporation of other components such as $Al_2O_3$ and the addition of the stabilizer $Y_2O_3$ in the LTD [41, 42].

Nevertheless, no comprehensive study has been conducted on the influence of microwave sintering on ATZ materials exposed to LTD environments. Since the sintering process and its associated conditions are a critical factor to determine the LTD behavior of 3Y-TZP-based materials, it is very relevant to research the effect of microwave sintering techniques on this behavior.

Presenda et al. [17] corroborate the high resistance to LTD of the microwave-sintered dense zirconia materials (3Y-TZP) as E and H remain almost unaltered; even after 140 h, E is still around 250 GPa and H is approximately 15 GPa. The starting material also has an important role in the microstructure evolution during sintering and, thus, in the LTD resistance. Microwave-sintered 3Y-TZP does not show any significant evidence of degradation after 200 h of exposure to LTD.

## 4. Conclusions

Microwave sintering has a very noticeable effect on the properties of ATZ material and therefore on their susceptibility to degradation at low temperatures. The ATZ composite has been obtained with densities close to theoretical densities by microwave heating, considerably reducing sintering times compared with the conventional sintering method. Microwave heating is accompanied by modification of the densification mechanisms, giving the ATZ material excellent mechanical properties.

In sample MW1300_10, the amount of the monoclinic phase is lower than for CS1500_120 after the same hours of exposure to LTD conditions. Therefore, the non-conventional fast technique of microwave sintering is a great alternative since it requires lower temperatures and times for densification, giving materials more resistant to low-temperature degradation.

The addition of alumina to the composition makes the phase change more difficult, the ATZ composite being more resistant to degradation than monolithic zirconia material.

In summary, there are many variables that can influence the resistance to LTD, including the composition of the material, quantity and distribution of the stabilizer or the sintering mechanisms among others. In this work, the results obtained indicate that both the addition of alumina in the composition and the use of microwaves as a sintering method favor the resistance to aging, occurring at a lower speed. There is no earlier publication on the effect of microwave sintering on the behavior of the LTD on ATZ materials, so this work is an approximation to understand this phenomenon.

## Acknowledgements

The authors gratefully acknowledge the funding support of the Spanish Ministry of Science and Innovation (RTI2018-099033-B-C32, RYC-2016-20915), Generalitat Valenciana (GRISOLIAP/2018/168) and Dr. Lorena Gil for contributing to the research work described in this chapter. We would also like to thank the researchers of the ITACA-UPV Institute for their contribution to the design and construction of the microwave equipment.

## Conflict of interest

The authors did not report any conflict of interest.

## Author details

Rut Benavente, Maria Dolores Salvador and Amparo Borrell*
Institute of Materials Technology, Polytechnic University of Valencia, Valencia, Spain

*Address all correspondence to: aborrell@upv.es

## References

[1] Kelly JR, Denry I. Stabilized zirconia as a structural ceramic: An overview. Dental Materials. 2008;**24**:289-298

[2] Sequeira S, Fernandes MH, Neves N, Almeida MM. Development and characterization of zirconia–alumina composites for orthopedic implants. Ceramics International. 2017;**43**:693-703

[3] Bona AD, Pecho OE, Alessandretti R. Zirconia as a dental biomaterial. Materials. 2015;**8**:4978-4991

[4] Li S, Izui H, Okano M, Zhang W, Watanabe T. Microstructure and mechanical properties of $ZrO_2$ ($Y_2O_3$)-$Al_2O_3$ nanocomposites prepared by spark plasma sintering. Particuology. 2012;**10**:345-351

[5] Meena KL, Karunakar DB. Development of alumina toughened zirconia nanocomposites using spark plasma sintering. Materials Today: Proceedings. 2018;**5**:16928-16935

[6] Torrecillas R, Moya JS, Diaz LA, Bartolomé JF, Fernández A, Lopez-esteban S. Nanotechnology in joint replacement. Wiley Interdisciplinary Reviews: Nanomedicine and Nanobiotechnology. 2009;**1**:540-552

[7] Faga MG, Vallée A, Bellosi A, Mazzocchi M, Thinh NN, Martra G, et al. Chemical treatment on alumina-zirconia composites inducing apatite formation with maintained mechanical properties. Journal of the European Ceramic Society. 2012;**32**:2113-2120

[8] Nevarez-Rascon A, Aguilar-Elguezabal A, Orrantia E, Bocanegra-Bernal MH. On the wide range of mechanical properties of ZTA and ATZ based dental ceramic composites by varying the $Al_2O_3$ and $ZrO_2$ content. International Journal of Refractory Metals and Hard Materials. 2009;**27**:962-970

[9] Benavente R, Salvador MD, Peñaranda FL, Pallone E, Borrell A. Mechanical properties and microstructural evolution of alumina-zirconia nanocomposites by microwave sintering. Ceramics International. 2014;**40**:11291-11297

[10] Fiedler S, Irsig R. Machining of biocompatible ceramics with femtosecond laser pulses. Biomed Technologies. 2013;**58**:1-2

[11] Sorrentino R, de Simone G, Tet S, Russo S, Zarone F. Five-year prospective clinical study of posterior three-unit zirconia-based fixed dental prostheses. Clinical Oral Investigations. 2012;**16**(3):977-985

[12] Lüthy H, Filser F, Loeffel O, Schumacher M, Gauckler LJ, Hammerle CHF. Strength and reliability of four-unit all-ceramic posterior bridges. Dental Materials. 2005;**21**:930-937

[13] Chevalier J, Gremillard L. Ceramics for medical applications: A picture for the next 20 years. Journal of the European Ceramic Society. 2009;**29**:1245-1255

[14] Kohorst P, Borchers L, Strempel J, Stiesch M, Hassel T, Bach FW, et al. Low-temperature degradation of different zirconia ceramics for dental applications. Acta Biomaterialia. 2012;**8**:1213-1220

[15] Ramesh S, Sara Lee KY, Tan CY. A review on the hydrothermal ageing behaviour of Y-TZP ceramics. Ceramics International. 2018;**44**:20620-20634

[16] Borrell A, Gil L, Presenda A, Salvador MD, Vleugels J, Moreno R. Influence of relative humidity and low temperature hydrothermal degradation on fretting wear of Y-TZP dental ceramics. Wear. 2019;**428-429**:1-9

[17] Presenda A, Salvador MD, Moreno R, Borrell A. Hydrothermal degradation behavior of Y-TZP ceramics sintered by nonconventional microwave technology. Journal of the American Ceramic Society. 2015;**98**:3680-3689

[18] Gi-Flores L, Salvador MD, Penaranda-Foix FL, Rosa R, Veronesi P, Leonelli C, et al. Low temperature degradation behaviour of 10Ce-TZP/Al2O3 bioceramics obtained by microwave sintering technology. In: AMPERE 2019-17th International Conference on Microwave and High Frequency Heating. ASPE: American Society for Precision Engineering; 2019. pp. 426-432

[19] Chevalier J, Gremillard L, Deville S. Low-temperature degradation of zirconia and implications for biomedical implants. Annual Review of Materials Research. 2007;**37**:1-32

[20] Norton MR, Yarlagadda R, Anderson GH. Catastrophic failure of the elite plus total hip replacement, with a hylamer acetabulum and zirconia ceramic femoral head. The Journal of Bone and Joint Surgery. British volume. 2002;**84**:631-635

[21] Mishra RR, Sharma AK. Microwave-material interaction phenomena: Heating mechanisms, challenges and opportunities in material processing. Composites Part A: Applied Science and Manufacturing. 2016;**81**:78-97

[22] Oghbaei M, Mirzaee O. Microwave versus conventional sintering: A review of fundamentals, advantages and applications. Journal of Alloys and Compounds. 2010;**494**:175-189

[23] Ramesh S, Zulkifli N, Tan CY, Wong YH, Tarlochan F, Ramesh S, et al. Comparison between microwave and conventional sintering on the properties and microstructural evolution of tetragonal zirconia. Ceramics International. 2018;**44**:8922-8927

[24] Gil-Flores L, Salvador MD, Penaranda-Foix FL, Fernández A, Suarez M, Rosa R, et al. Microstructure and mechanical properties of 5.8 GHz microwave-sintered $ZrO_2/Al_2O_3$ ceramics. Ceramics International. 2019;**45**:18059-18064

[25] Inokoshi M, Zhang F, De Munck J, Minakuchi S, Naert I, Vleugels J, et al. Influence of sintering conditions on low-temperature degradation of dental zirconia. Dental Materials. 2014;**30**:669-678

[26] Zhang F, Vanmeensel K, Inokoshi M, Batuk M, Hadermann J, Van Meerbeek B, et al. Critical influence of alumina content on the low temperature degradation of 2-3 Mol% yttria-stabilized TZP for dental restorations. Journal of the European Ceramic Society. 2015;**35**:741-750

[27] Guillén RM, Salvador MD, Gutiérrez-González CF, Catalá-Civera JM, Borrell A. Fabrication and characterization of $Nb_2O_5$-doped 3Y-TZP materials sintered by microwave technology. International Journal of Applied Ceramic Technology. 2021;**18**:2033-2044

[28] Zhang M, Bao X, Barmuta P, Bao J, Markovic T, Ocket I, et al. Fully automated electrically controlled tunable broadband interferometric dielectric spectroscopy for aqueous solutions. IEEE Transactions on Microwave Theory and Techniques. 2022;**70**:532-541

[29] Huang Z, Gotoh M, Hirose Y. Improving sinterability of ceramics using hybrid microwave heating. Journal of Materials Processing Technology. 2009;**209**:2446-2452

[30] Borrell A, Salvador MD, Miranda M, et al. Microwave technique: A powerful tool for sintering ceramic materials. Current Nanoscience. 2014;**10**:32-35

[31] Gil-Flores L, Salvador MD, Penaranda-Foix FL, Dalmau A, Fernández A, Borrell A. Tribological and wear behaviour of alumina toughened zirconia nanocomposites obtained by pressureless rapid microwave sintering. Journal of the Mechanical Behavior of Biomedical Materials. 2020;**101**:103415

[32] Borrell A, Salvador MD, Penaranda-Foix FL, Catalá-Civera JM. Microwave sintering of zirconia materials: Mechanical and microstructural properties. International Journal of Applied Ceramic Technology. 2013;**10**:313-320

[33] Rayón E, Moreno R, Alcazar C, Salvador MD, Manjón FJ, Jiménez-Piqué E, et al. Enhanced hydrothermal resistance of Y-TZP ceramics through colloidal processing. Journal of the American Ceramic Society. 2013;**96**:1070-1076

[34] Lim CS, Finlayson TR, Ninio F, Griffiths JR. In-situ measurement of the stress-induced phase transformations in magnesia-partially-stabilized zirconia using Raman spectroscopy. Journal of the American Ceramic Society. 1992;**75**:1570-1573

[35] Muñoz Tabares JA, Anglada MJ. Quantitative analysis of monoclinic phase in 3Y-TZP by raman spectroscopy. Journal of the American Ceramic Society. 2010;**93**:1790-1795

[36] Oliver WC, Pharr GM. An improved technique for determining hardness and elastic modulus using load and displacement sensing indentation experiments. Journal of Materials Research. 1992;**19**:1564-1583

[37] Wu ZK, Li N, Jian C, Zhao W-Q, Yan J-Z. Low temperature degradation of $Al_2O_3$-doped 3Y-TZP sintered at various temperatures. Ceramics International. 2013;**39**:7199-7204

[38] Presenda A, Salvador MD, Penaranda-Foix FL, Moreno R, Borrell A. Effect of microwave sintering on microstructure and mechanical properties in Y-TZP materials used for dental applications. Ceramics International. 2015;**41**:7125-7132

[39] Presenda A, Salvador MD, Penaranda-Foix FL, Catalá-Civera JM, Pallone E, Ferreira J, et al. Effects of microwave sintering in aging resistance of zirconia-based ceramics. Chemical Engineering and Processing. 2017;**122**:404-412

[40] Chevalier J, Gremillard L, Deville S. Low-temperature degradation of zirconia and implications for biomedical implants. Annual Review of Materials Research. 2007;**37**:1-32

[41] Inokoshi M, Zhang F, Vleugels J, Vanmeensel K. Influence of sintering conditions on low-temperature degradation of dental zirconia. Dental Materials. 2014;**30**:669-678

[42] Zhang F, Vanmeensel K, Inokoshi M, Batuk M, Hadermann J, Van Meerbeek B, et al. Critical influence of alumina content on the low temperature degradation of 2-3 Mol% Yttria-stabilized TZP for dental restorations. Journal of the European Ceramic Society. 2015;**35**:741-750

Chapter 6

# Synthesis and Investigation of Ceramic Materials for Medium-Temperature Solid Oxide Fuel Cells

*Marina V. Kalinina, Daria A. Dyuskina, Irina G. Polyakova, Sergey V. Mjakin, Maxim Yu. Arsent'ev and Olga A. Shilova*

## Abstract

Finely dispersed $(CeO_2)_{1-x}(Sm_2O_3)_x$ (x = 0.02; 0.05; 0.10); $La_{1-x}Sr_xNiO_3$, $La_{1-x}Sr_xCoO_3$ and $La_{1-x}Sr_xFe_{0.7}Ni_{0.3}O_3$ (x = 0.30; 0.40) mesoporous xerogel powders are synthesized by co-crystallization of the corresponding nitrates with ultrasonic processing and used to obtain nanoscale ceramic materials with cubic fluorite-like, orthorhombic, and perovskite-like tetragonal crystal structure, respectively, with CSR ~ 64–81 nm (1300°C). Physicochemical characterization of the obtained ceramics revealed that $(CeO_2)_{1-x}(Sm_2O_3)_x$ features with open porosity 2–6%, while for $La_{1-x}Sr_xNiO_3$, $La_{1-x}Sr_xCoO_3$, and $La_{1-x}Sr_xFe_{0.7}Ni_{0.3}O_3$, this value is about 21–29%. Ceria-based materials possess a predominantly ionic conductivity (ion transport numbers $t_i$ = 0.82–0.71 in the temperature range 300–700°C, $\sigma_{700°C} = 1.3\cdot10^{-2}$ S/cm) determined by the formation of mobile oxygen vacancies upon heterovalent substitution of $Sm^{3+}$ for $Ce^{4+}$. For solid solutions based on lanthanum nickelate and cobaltite, a mixed electronic-ionic conductivity ($\sigma_{700°C} = 0.80\cdot10^{-1}$ S/cm) with ion transport numbers ($t_e$ = 0.98–0.90, $t_i$ = 0.02–0.10) was obtained. The obtained ceramic materials are shown to be promising as solid oxide electrolytes and electrodes for medium-temperature fuel cells.

**Keywords:** co-crystallization of salts, oxides, finely dispersed powders, fuel cells, nanoceramics, electrolytes, electrode materials

## 1. Introduction

In view of the currently increasing demand for energy resources, gradual depletion of fossil fuels, and growing environmental problems, the development of alternative hydrogen energy is an essential and highly promising R&D direction. For further progress in this area, the development of advanced materials for electrochemical power systems is required. Particularly, promising solid oxide fuel cells (SOFCs) provide the efficiency of up to 85% in couple with almost double cost savings and a 100 times reduction of harmful emissions compared with conventional power sources due to the absence of direct chemical contact between the fuel and the oxidizer [1]. The application of SOFC-based power sources a significant power and fuel savings. The operation temperatures as high as 700–950°C afford increased rates of electrode reactions without using expensive catalysts. Another advantage of SOFCs is the absence of strict requirements for fuel purity. Besides hydrogen, any hydrocarbons converted into synthesis gas ($H_2$-CO) can be used. High thermodynamic efficiency, continuous operation, and environment-friendly performances make solid oxide fuel cells more promising compared with such conventional systems as internal combustion engines, solar panels, and wind turbines. As a promising power storage system, reversible SOFCs can provide an economically effective approach to the power management using discontinuous energy sources [2, 3].

Thus, the development and implementation of SOFC-based fuel cells for commercial production are becoming a priority to address the problems of distributed power supply, energy saving, cogeneration, and saving fuel resources.

The decrease of working temperature and development of medium-temperature SOFCs is an important goal in materials science since high-operating temperatures cause problems with the compatibility of electrode materials and electrolytes. The implementation of medium-temperature SOFCs can provide an extension of the range of applied materials, reduce degradation of the devices, and increase their operational lifetime. The main components of SOFC include cathode, anode, and electrolyte. The applied electrolytes differ in their anionic, protonic, or ion-mixed ion-transport mechanism. The basic principle of fuel cell operation is that the transport of oxygen ions ($O^{2-}$) from the cathode to the anode can only proceed in the presence of oxygen vacancies. In this regard, the optimal electrolyte materials must contain anion vacancies in the crystal lattice. Currently, cerium dioxide ($CeO_2$)-based nanomaterials with oxygen-ionic conductivity are considered promising as medium-temperature electrolytes affording the reduction of the fuel cell operating temperature by 300–400°C. In respect of electric performances, these electrolytes are not inferior to conventional zirconia-based YSZ materials (particularly, $(ZrO_2)_{0.92}(Y_2O_3)_{0.08}$ ceramics) [4, 5]. Furthermore, they are thermodynamically stable at relatively low-operating temperatures of 600–800°C that provide a long lifetime. Moreover, relatively low-operating temperatures prevent from the interlayer diffusion and interfacial layers between SOFC components exclude solid-phase interaction. All the existing SOFC cathodes have certain disadvantages that determine the growing interest in R&D in the development of new cathode materials, particularly for medium-temperature (500–700°C) SOFCs.

During SOFC operation in the middle-temperature range, the power characteristics of the cell are limited by the cathode operation [6]. Since the SOFC cathode reduction of molecular oxygen and transport of oxygen ions to the electrolyte surface take place, the developed medium-temperature cathode materials should meet several requirements. Particularly, they should be ultrafine and have high electronic or mixed electron-ion conductivity. To reduce the diffusion drag at the cathode, a well-developed porous structure is required, since the rate of molecular oxygen reduction depends on the specific surface area. Furthermore, the mechanical compatibility of the cathode and electrolyte is important. These requirements are fulfilled by complex metal oxides and their nanocomposites, particularly complex oxides based on rare earth elements (REE) and 3D transition metals combining high electric and catalytic properties. Currently, single-phase complex perovskite materials such as $LnMO_{3-\delta}$ (M-Cr, Mn, Fe, Ni, and Co) are proposed as new materials for medium-temperature SOFC cathodes, featuring stability in an oxidizing atmosphere in a wide temperature range and sufficiently high p-type electrical conductivity [6, 7]. Among them, the highest conductivity is observed for manganese-, cobalt-, and nickel-containing oxides. The conductivity of such perovskites can be further enhanced by increasing the concentration of charge carriers (holes) due to the heterovalent substitution of $La^{3+}$ for cations of alkaline earth elements $M^{2+}$ = Ca, Sr, or Ba [8]. Oxides with a perovskite structure constitute a large class of complex oxides with $ABO_3$-type unit cells (**Figure 1**). A distinctive feature of perovskites is the possibility of cationic substitution in both A and B positions in a wide range of concentrations [9]. In practice, most crystals with a cubic perovskite structure crystallize in a lower symmetry with the distortion of the cubic structure to orthorhombic, hexagonal, or tetragonal one.

The variation of the substituting cations percentage within a fairly wide range and changing their oxidation degrees allow the simulation of functional properties of perovskite-like oxide systems.

Strontium lanthanum manganites such as $La_{1-x}Sr_xMnO_3$ (LSM) with perovskite structure are extensively used as cathodes of solid oxide fuel cells, providing the best performances in the high-temperature range (800–1000°C) [10]. For medium-temperature fuel cells, lanthanum cobaltites and nickelates seem to be most promising, since their electrical conductivity exceeds that of lanthanum manganites due to a higher specific surface reactivity determined by a lower strength of metal-oxygen (Me–O) bond [11]. The development of processes for obtaining efficient electrolyte and cathode materials for SOFCs is an important R&D goal.

The most economically efficient approach to this problem is the use of liquid-phase synthetic methods, including co-crystallization of salts, co-precipitation of hydroxides, sol-gel, and hydrothermal. These approaches afford fine powders and nanoceramic materials-based thereon, also providing a reduced energy consumption due to the reduced temperature of powder synthesis and ceramics sintering [12].

The preparation of solid oxide electrolyte and cathode materials and their characterization to find the relationships such as "composition – synthesis technology – structure – properties" afford the determination of their optimal characteristics and synthesis conditions.

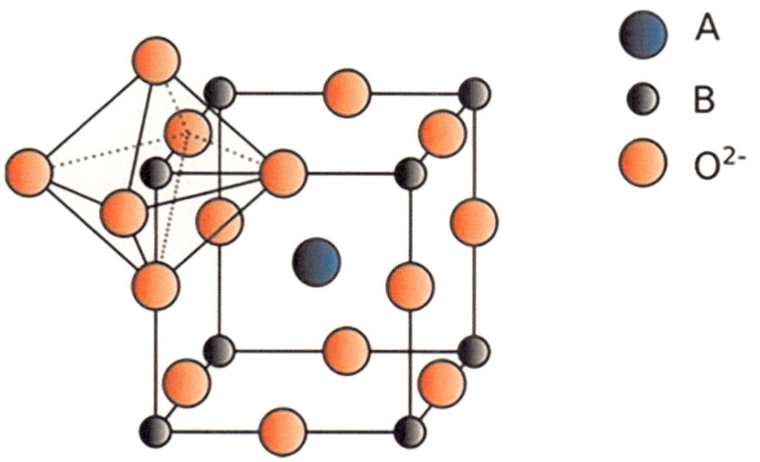


**Figure 1.**
*Perovskite structure.*

## 2. Synthesis and investigation of ceramic materials

### 2.1 Synthesis of ceramic materials

The synthesis of xerogels and nanodispersed powders with different oxide concentration ratios in the $CeO_2$–$Sm_2O_3$, $La_2O_3$–SrO–$Ni_2O_3$($Co_2O_3$), and $La_2O_3$–SrO–$Ni_2O_3$–$Fe_2O_3$ systems was carried out by co-crystallization of the corresponding nitrates followed by the ultrasonic treatment [13]. Nitrates of cerium $Ce(NO_3)_3{\cdot}6H_2O$ (analytical grade), samarium $Sm(NO_3)_3{\cdot}6H_2O$ (chemically pure), lanthanum La $(NO_3)_3{\cdot}6H_2O$ (chemically pure), strontium Sr.$(NO_3)_2$ (analytical grade), nickel Ni $(NO_3)_2{\cdot}6H_2O$ (pure), cobalt $Co(NO_3)_2{\cdot}6H_2O$ (pure), and iron $Fe(NO_3)_3{\cdot}9H_2O$ (pure) as ~0.5 M solutions were used for the synthesis. The resulting solutions were mixed in the proportions corresponding to the stoichiometric ratio of oxides, followed by evaporation on a water bath for 3 h to obtain a supersaturated solution. Then, the supersaturated solution was cooled at a temperature of 3–5°C to facilitate the adsorption of the crystallizing substance on the surface of the crystals formed during the evaporation of mixtures of the nitrate solutions. To improve the dispersity of crystalline particles and to make their size distribution more narrow, the crystalline hydrate was subjected to ultrasonic treatment for 30 min in distilled water, resulting in an almost monodispersed powder.

The subsequent drying (110°C, 0.5 h) yielded X-ray amorphous xerogels that were subjected to heat treatment (600°C, 1 h) to form nanopowders with a stable crystal structure. The synthesized powders of a given composition were ground in a mortar, followed by uniaxial cold pressing into tablets of 1.0 and 1.5 cm in diameter using a PGR400 at the pressure of 100–150 MPa and then sintering installation for 2 hours and temperature 1300°C.

### 2.2 Characterization techniques

XRD characterization was performed using a Bruker D8-Advanced diffractometer. The ICDD-2006 international database was used to interpret the diffraction patterns, and the analysis results were processed using the WINFIT 1.2.1 software involving the Fourier transform of the reflection profile. The sizes of coherent scattering regions (CSR) were estimated according to the Selyakov-Scherer equation:

$$D_{CSR} = \frac{0.9 \times \lambda}{\beta \times \cos\theta} \quad (1)$$

where $\lambda$ = 15,418 Å is the CuKα wavelength, and $\beta$ is the XRD peak full width at half maximum (FWHM) [14].

The thermolysis processes occurring in coprecipitated xerogels and powders upon heating in the temperature range of 20–1000°C were studied using the Q-1000 derivatograph (MOM). The specific surface area of the synthesized nanopowders was measured by low-temperature nitrogen adsorption using a QuantaChrome Nova 4200B analyzer. Based on the obtained data, the specific surface area $S_{BET}$ of the samples was calculated using the Brunauer-Emmett-Teller (BET) model. The calculation of the pore size distribution was carried out based on nitrogen desorption isotherms according to the Barret-Joyner-Halenda (BJH) method; the thermal treatment of the powders was carried out in a Naberterm furnace with program control in the temperature range of 25–1300°C for 3 hours, followed by slow cooling of the furnace.

The open porosity of the samples was determined by hydrostatic weighing in distilled water in accordance with the Russian standard GOST 473.4–81 [15]. The electrical resistance of the obtained ceramic materials was measured by the two-contact method at a direct current in the temperature range of 250–1000°C using the "Hardware-software complex for studying the electrical properties of nanoceramics in various gaseous media" [16].

The transfer numbers of ions and electrons in bulk solid electrolytes [17] were determined using the West-Tallan method. A mixture of $CO_2$ + CO was used as an inert gas (corresponding to a partial pressure of oxygen $10^3$ Pa). The measurements were carried out at a direct current in weak (U = 0.5 V) fields after a long (up to 30 min) current drop. The contributions of ionic and electronic conductivities were estimated according to the formulas:

$$t_e = \frac{R_{air}}{R_e} \tag{2}$$

$$t_i = 1 - t_e \tag{3}$$

where $t_e$ and $t_i$ are electron and ion transport numbers, respectively; $R_{air}$ and $R_e$ are electric resistance values measured in air and inert gas, respectively.

### 2.3 Results and discussion

For all the obtained compositions, thermolysis processes were studied. As an example, DTA thermograms of $La_{0.6}Sr_{0.4}CoO_3$ xerogel synthesized with (a) and without ultrasonic treatment (b) are shown in **Figure 2**.

As shown in **Figure 2**, ultrasonic treatment during the synthesis results in the reduction of the crystalline hydrate dehydration temperature from 115–110°C, as well as the temperatures of all thermal transformations. This effect is determined by weakening the bonds between the molecules of nitrate salts and crystallization water molecules upon the impact of ultrasonic waves, facilitating the dehydration and decomposition of salts. Ultrasonic treatment also affects the powder crystallization reducing the temperature of its transition into the crystalline phase (320 → 290°C).

Compared with the considered data, the differences in the temperatures of exo- and endo-effects for other compositions were no more than 10–15°C.

The microstructural parameters of the synthesized powders were determined using low-temperature nitrogen adsorption.

Microstructural performances of the synthesized powders are summarized in **Table 1**.

XRD characterization revealed that at 600°C a cubic solid solution of the fluorite type is formed in the studied $CeO_2$–$Sm_2O_3$ powders with an average CSR size of ~10 nm. Subsequent annealing at higher temperatures (1300°C) does not disrupt the single-phase structure of the nanopowders and ceramics-based thereon. As an example, consecutive steps of fluorite-type cubic solid solution formation for the $(CeO_2)_{0.95}(Sm_2O_3)_{0.05}$ sample are shown in **Figure 3**. Crystal structures and specific electrical conductivity of synthesized powders and ceramics in the system $CeO_2$–$Sm_2O_3$ are shown in **Table 2**.

The phase compositions of powders and corresponding ceramics of all the compositions in the systems $La_2O_3$–SrO–$Ni_2O_3$($Co_2O_3$) and $La_2O_3$–SrO–$Ni_2O_3$–$Fe_2O_3$ are

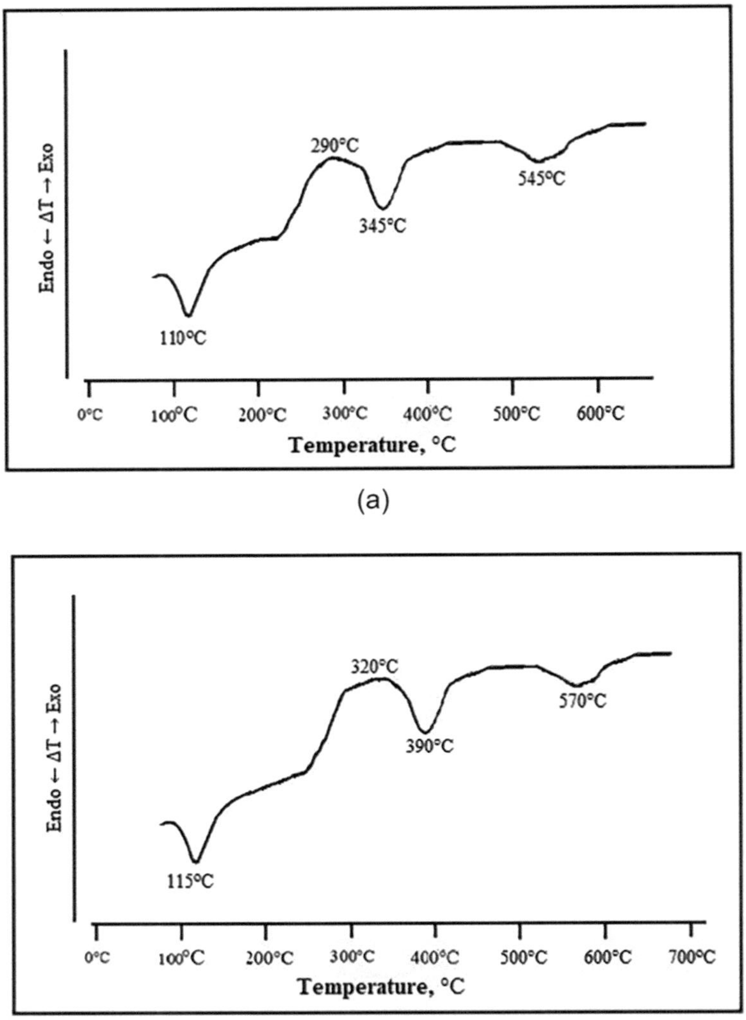


**Figure 2.**
*Differential thermal analysis results for $La_{0.6}Sr_{0.4}CoO_3$ xerogel prepared with (a) and without (b) xerogel freezing at −25°C (24 h).*

| Composition | Specific surface area Ss, $m^2/g$ | Average pore size $D_{por}$, nm | Specific pore volume $V_{por}$, $cm^3/g$ |
|---|---|---|---|
| Electrolyte materials | | | |
| $(CeO_2)_{0.98}(Sm_2O_3)_{0.02}$ | 40 | 11 | 0.075 |
| $(CeO_2)_{0.95}(Sm_2O_3)_{0.05}$ | 63 | 8 | 0.082 |
| $(CeO_2)_{0.90}(Sm_2O_3)_{0.10}$ | 74 | 2 | 0.087 |
| Cathode materials | | | |
| $La_{0.7}Sr_{0.3}NiO_3$ | 30 | 2.3 | 0.029 |
| $La_{0.6}Sr_{0.4}NiO_3$ | 28 | 2.4 | 0.026 |

| Composition | Specific surface area Ss, $m^2/g$ | Average pore size $D_{por}$, nm | Specific pore volume $V_{por}$, $cm^3/g$ |
|---|---|---|---|
| $La_{0.7}Sr_{0.3}CoO_3$ | 15 | 2.1 | 0.018 |
| $La_{0.6}Sr_{0.4}CoO_3$ | 15 | 2.2 | 0.020 |
| $La_{0.7}Sr_{0.3}Fe_{0.7}Ni_{0.3}O_3$ | 31 | 2.2 | 0.035 |
| $La_{0.6}Sr_{0.4}Fe_{0.7}Ni_{0.3}O_3$ | 30 | 2.0 | 0.033 |

**Table 1.**
*Microstructural performances of the synthesized powders [18].*

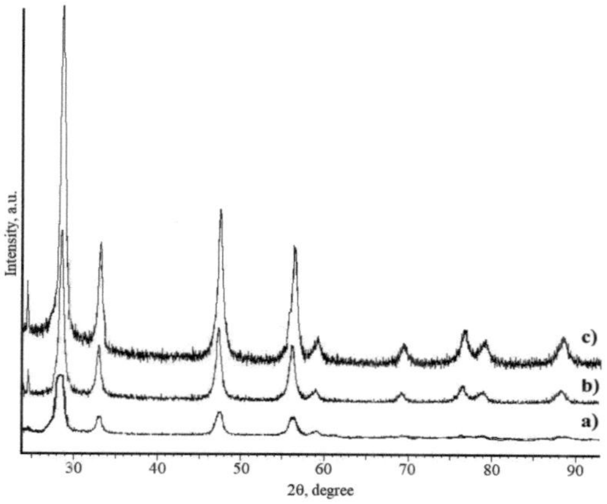


**Figure 3.**
*XRD patterns of $(CeO_2)_{0.90}(Sm_2O_3)_{0.10}$ (a = 5.41396 Å) based xerogel (a—150°C), nanopowder (b—600°C), and ceramics (c—1300°C) samples [18].*

| Composition | Fluorite-like structure | Lattice parameter, a (Å) | CSR, nm | | $\sigma \cdot 10^{-2}$, $S \cdot cm^{-1}$ (700°C) | $E_a$, eV |
|---|---|---|---|---|---|---|
| | | | 600°C | 1300°C | | |
| $(CeO_2)_{0.98}(Sm_2O_3)_{0.02}$ | cubic | a = 5.43204 | 11 | 81 | 0.4 | 1.15 |
| $(CeO_2)_{0.95}(Sm_2O_3)_{0.05}$ | cubic | a = 5.41848 | 9 | 70 | 0.7 | 1.13 |
| $(CeO_2)_{0.90}(Sm_2O_3)_{0.10}$ | cubic | a = 5.41396 | 8 | 68 | 1.3 | 1.00 |

**Table 2.**
*Crystal structures and specific electrical conductivity of powders and ceramics in the system $CeO_2$–$Sm_2O_3$ [18].*

summarized in **Table 3**. As an example, **Figure 4** shows the phase composition of $La_{0.6}Sr_{0.4}NiO_3$ xerogel powder and ceramics, indicating the formation of a solid solution with a tetragonal perovskite structure at 1300°C.

The data in **Table 3** show that the synthesized powders and ceramic materials in the temperature range of 600–1300°C have an orthorhombic and tetragonal perovskite-type structure.

Physicochemical properties of all the ceramic samples obtained on the basis of nanopowders in the $CeO_2$-$Sm_2O_3$ $La_2O_3$–SrO–$Ni_2O_3$($Co_2O_3$) and $La_2O_3$–SrO–$Ni_2O_3$–$Fe_2O_3$ systems are summarized in **Table 4**. These data show that an increase in the content of samarium oxide in the obtained samples in the $CeO_2$-$Sm_2O_3$ system leads to a decrease in their density, which is probably due to the distortion of the cerium dioxide lattice upon $Sm_2O_3$ dissolving.

To create a porous structure in ceramics based on the $La_2O_3$–SrO–$Ni_2O_3$($Co_2O_3$) and $La_2O_3$–SrO–$Ni_2O_3$–$Fe_2O_3$ systems, a pore-forming additive (10% aqueous solution of

| Composition (preparation conditions) | Perovskite-like structure | Parameters | CSR, nm | $\sigma \cdot 10^{-1}$, $S \cdot cm^{-1}$ (700°C) | $E_a$, eV |
|---|---|---|---|---|---|
| $La_{1-x}Sr_xCoO_3$ | | | | | |
| $La_{0.7}Sr_{0.3}CoO_3$ (600°C, holding 1 h) | Orthorhombic (crystallization start) | a = 4.1132<br>b = 5.429<br>c = 9.804<br>V = 218.93 | 34 | Non-sintered powder | — |
| $La_{0.7}Sr_{0.3}CoO_3$ (1300°C, heating 3 h + holding 5 h) | Orthorhombic | a = 4.1027<br>b = 5.420<br>c = 9.792<br>V = 217.75 | 70 | Ceramics 0.20 | 1.82 |
| $La_{0.6}Sr_{0.4}CoO_3$ (1300°C, heating 3 h + holding 5 h) | Tetragonal | a = 3.9052<br>c = 12.8412<br>V = 195.84 | 67 | Ceramics 0.35 | 1.67 |
| $La_{1-x}Sr_xNiO_3$ | | | | | |
| $La_{0.7}Sr_{0.3}NiO_3$ (900°C, holding 3 h) | Tetragonal | a = 3.8168<br>c = 12.7498<br>V = 185.74 | 40 | Non-sintered powder | — |
| $La_{0.7}Sr_{0.3}NiO_3$ (1300°C, heating 3 h + holding 2 h) | Tetragonal | a = 3.8140<br>c = 12.7421<br>V = 185.35 | 68 | Ceramics 0.25 | 1.78 |
| $La_{0.6}Sr_{0.4}NiO_3$ (600°C, holding 1 h) | Orthorhombic | a = 14.6033<br>b = 9.7345<br>c = 6.9057<br>V = 981.69 | 32 | Non-sintered powder | — |
| $La_{0.6}Sr_{0.4}NiO_3$ (900°C, holding 3 h) | Tetragonal | a = 3.8187<br>c = 12.7496<br>V = 185.92 | 44 | Non-sintered powder | — |
| $La_{0.6}Sr_{0.4}NiO_3$ (1300°C, heating 3 h + holding 2 h) | Tetragonal | a = 3.8152<br>c = 12.7436<br>V = 185.49 | 65 | Ceramics 0.80 | 1.63 |
| $La_{1-x}Sr_xFe_{1-y}Ni_yO_3$ | | | | | |
| $La_{0.6}Sr_{0.4}Fe_{0.7}Ni_{0.3}O_3$ (600°C, holding 1 h) | Orthorhombic | a = 15.10<br>b = 9.687<br>c = 6.9297<br>V = 1013.63 | 25 | Non-sintered powder | — |
| $La_{0.6}Sr_{0.4}Fe_{0.7}Ni_{0.3}O_3$ (1300°C, heating 3 h + holding 2 h) | Tetragonal | a = 3.9102<br>c = 12.7903<br>V = 195.33 | 65 | Ceramics 0.75 | 1.59 |
| $La_{0.7}Sr_{0.3}Fe_{0.7}Ni_{0.3}O_3$ (1300°C, heating 3 h + holding 2 h) | Tetragonal | a = 3.8965<br>c = 12.7752<br>V = 193.96 | 63 | Ceramics 0.30 | 1.84 |

**Table 3.**
*Crystal structures and specific electrical conductivity of powders and ceramics in the systems $La_2O_3$–SrO–$Ni_2O_3$($Co_2O_3$) and $La_2O_3$–SrO–$Ni_2O_3$–$Fe_2O_3$.*

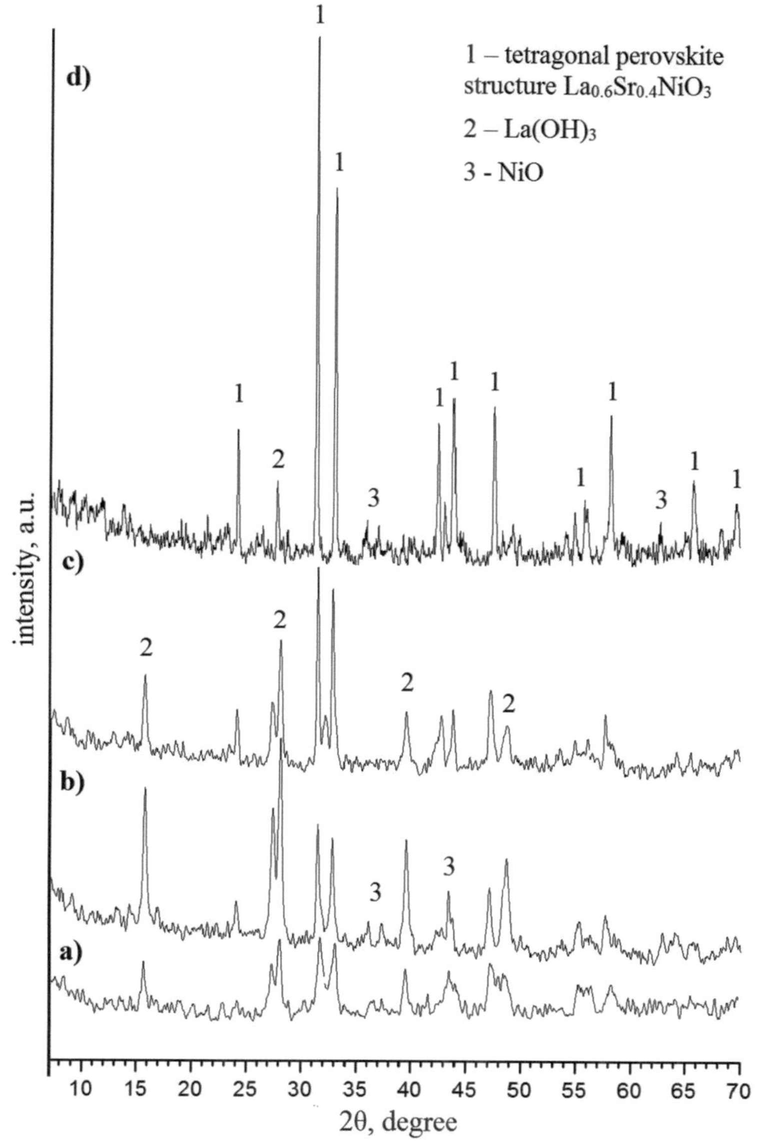


**Figure 4.**
*XRD patterns of $La_{0.6}Sr_{0.4}NiO_3$ xerogel powder and ceramics; heat treatment at a) 600°C, b) 900°C, c) 1000°C, and d) 1300°C.*

polyvinyl alcohol) in an amount of 10% over the bulk of the charge was added [19]. The values of open porosity are in the range of 21–29%, which is one of the main conditions for the optimal operation of cathode solid oxide materials as shown in **Table 4**.

One of the main conditions for SOFC operation is the compatibility of its components in terms of the thermal expansion coefficient (TEC). TEC of electrolytes based on cerium oxide doped with samarium oxide is $\alpha = 12.5 \cdot 10^{-6}\ K^{-1}$ [20]. Cathode materials based on systems $La_2O_3$–SrO–$Ni_2O_3$ (TEC—$14.2 \cdot 10^{-6}\ K^{-1}$) and $La_2O_3$–SrO–$Ni_2O_3$–$Fe_2O_3$ (TEC—12.8–$13.1 \cdot 10^{-6}\ K^{-1}$) have comparable TEC values with ones of electrolytes based on cerium oxide. As can be seen in **Table 4**, strontium

| Composition | ΔL/L, % | $\rho_{teor}$, g/cm$^3$ | Apparent density $\rho_{exp}$, g/cm$^3$ | P, % | TEC α, ×10$^{-6}$ K$^{-1}$ |
|---|---|---|---|---|---|
| Electrolyte materials | | | | | |
| $(CeO_2)_{0.98}(Sm_2O_3)_{0.02}$ | 15.8 | 7.15 | 6.35 | 4.7 | 12.5 [20] |
| $(CeO_2)_{0.95}(Sm_2O_3)_{0.05}$ | 15.0 | 6.92 | 6.21 | 2.3 | 12.5 [20] |
| $(CeO_2)_{0.90}(Sm_2O_3)_{0.10}$ | 14.5 | 6.83 | 6.01 | 6.0 | 12.5 [20] |
| Cathode materials | | | | | |
| $La_{0.6}Sr_{0.4}NiO_3$ | 15.1 | 5.67 | 5.41 | 25.2 | 14.2 [21] |
| $La_{0.6}Sr_{0.4}CoO_3$ | 14.2 | 5.87 | 5.61 | 26.0 | 18.9 [22] |
| $La_{0.7}Sr_{0.3}NiO_3$ | 15.1 | 5.56 | 5.24 | 21.0 | 14.2 [21] |
| $La_{0.7}Sr_{0.3}CoO_3$ | 14.3 | 5.89 | 5.61 | 23.5 | 17.7 [22] |
| $La_{0.6}Sr_{0.4}Fe_{0.7}Ni_{0.3}O_3$ | 14.7 | 5.69 | 5.49 | 28.3 | 12.8–13.1 [20] |
| $La_{0.7}Sr_{0.3}Fe_{0.7}Ni_{0.3}O_3$ | 14.6 | 5.73 | 5.52 | 27.9 | 12.8–13.1 [20] |

**Table 4.**
*Physicochemical properties of ceramics samples synthesized by co-precipitation of salts [18].*

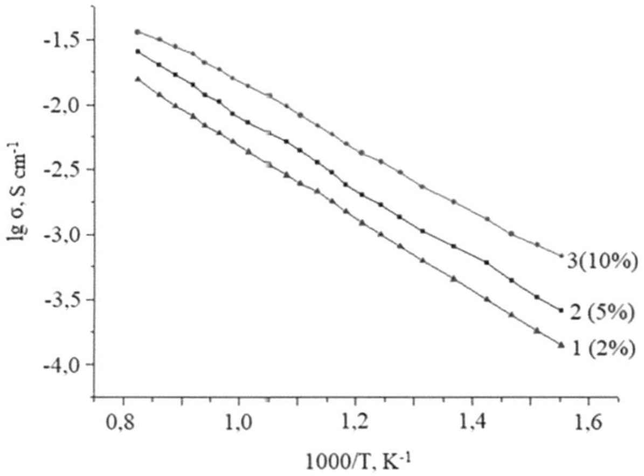


**Figure 5.**
*Temperature dependence for specific electrical conductivity of 1—$(CeO_2)_{0.95}(Sm_2O_3)_{0.02}$, 2—$(CeO_2)_{0.90}(Sm_2O_3)_{0.05}$, 3—$(CeO_2)_{0.80}(Sm_2O_3)_{0.10}$ [18].*

lanthanum cobaltites have a TEC much higher than the TEC of samples of the composition $(CeO_2)_{1-x}(Sm_2O_3)_x$.

The electrical conductivity of $(CeO_2)_{1-x}(Sm_2O_3)_x$ samples (x = 0.02; 0.05; 0.10) was measured using the two-contact method at direct current (**Figure 5**). The appearance of high oxygen ionic conductivity in $CeO_2$-$Sm_2O_3$-based solid electrolytes is determined by the formation of oxygen vacancies in the $CeO_2$ matrix when $Ce^{4+}$ is replaced by $Sm^{3+}$ during the dissolution of $Sm_2O_3$ in $CeO_2$, which can be described by the following quasi-chemical equation in the Kroeger-Winke notation [23]:

$$Sm_2O_3 \xrightarrow{CeO_2} 2Sm'_{Ce} + 3O_O^x + V_O^{\bullet\bullet} \quad (4)$$

where $Sm'_{Ce}$ is a samarium ion replacing $Ce^{4+}$ and yielding a negative charge, $V_O^{\bullet\bullet}$ is a positively charged oxygen vacancy compensating the dopant charge, and $O_O^{\times}$ is oxygen atom in a regular site with a neutral charge.

As can be seen in **Figure 5**, the temperature growth in the range from 500 to 1000°C leads to the increase in electrical conductivity of all the samples. In addition, with an increase in the concentration of samarium oxide, the specific electrical conductivity of the ceramics increases in the entire temperature range in the study. The highest specific electrical conductivity in the temperature range of 500–1000°C ($\sigma_{700°C} = 1.3 \cdot 10^{-2}$ S/cm) is observed for the sample containing 10 mol. % $Sm_2O_3$.

The temperature dependence of the specific electrical conductivity of solid solutions in the systems $La_2O_3$–SrO–$Co_2O_3$ and $La_2O_3$–SrO–$Ni_2O_3$ ($La_{0.6}Sr_{0.4}CoO_3$, $La_{0.7}Sr_{0.3}CoO_3$, $La_{0.6}Sr_{0.4}NiO_3$, $La_{0.7}Sr_{0.3}NiO_3$) is shown in **Figure 6**. The conductivity grows with temperature in the range from 300 to 700°C up to the saturation plateau at 600–800°C. Particularly, for $La_{0.6}Sr_{0.4}NiO_3$ and $La_{0.6}Sr_{0.4}CoO_3$, the plateau is reached at 600 and 700°C, respectively, while at higher temperatures, no conductivity increase is observed due to a change in the conductivity mechanism from semiconductor to metallic. The electrical conductivity in the considered solid solutions can proceed *via* several possible mechanisms, mainly by itinerant electrons along the $Ni^{3+}$–O–$Ni^{3+}$ chain and electron or hole jump directly between $Ni^{3+}$ and $Ni^{2+}$ ions. The authors of ref. [6] believe that the appearance of metallic conductivity in the synthesized solid solutions is due to the delocalization of Ni d-electrons during the interaction of nickel and oxygen atoms in the Ni-O-Ni chains.

**Figure 6** also indicates that the conductivity of the studied samples grows with an increase in strontium oxide content in the solid solutions.

The highest conductivity in the temperature range 500–1000°C ($\sigma_{700°C} = 0.80 \cdot 10^{-1}$ S/cm) is observed for the composition $La_{0.6}Sr_{0.4}NiO_3$. The conductivity values $\sigma_{700°C}$ for $La_{0.7}Sr_{0.3}NiO_3$, $La_{0.6}Sr_{0.4}CoO_3$, and $La_{0.7}Sr_{0.3}CoO_3$ samples are $0.25 \cdot 10^{-1}$, $0.35 \cdot 10^{-1}$, and $0.20 \cdot 10^{-1}$ S/cm, respectively. **Figure 7** shows the temperature dependence plots for samples of the compositions $La_{0.6}Sr_{0.4}Fe_{0.7}Ni_{0.3}O_3$ and $La_{0.7}Sr_{0.3}Fe_{0.7}Ni_{0.3}O_3$, indicating that the former one features with higher

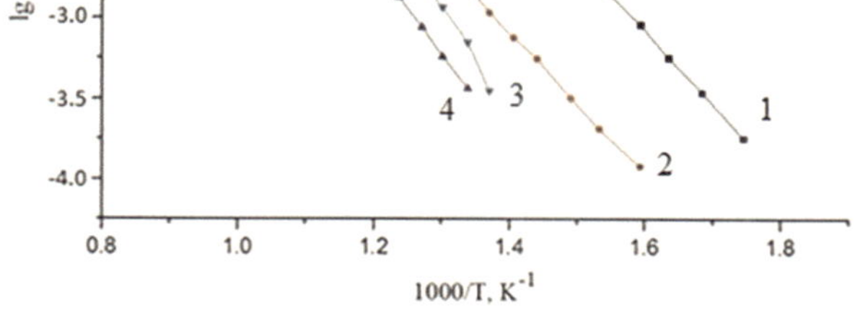


**Figure 6.**
*Temperature dependence for specific electrical conductivity of 1—$La_{0.6}Sr_{0.4}NiO_3$, 2—$La_{0.6}Sr_{0.4}CoO_3$, 3—$La_{0.7}Sr_{0.3}NiO_3$, and 4—$La_{0.7}Sr_{0.3}CoO_3$.*

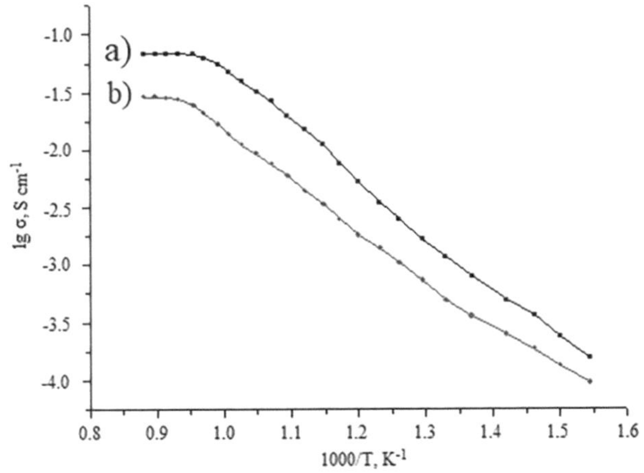


**Figure 7.**
*Temperature dependence for specific electrical conductivity of a) $La_{0.6}Sr_{0.4}Fe_{0.7}Ni_{0.3}O_3$ and b) $La_{0.7}Sr_{0.3}Fe_{0.7}Ni_{0.3}O_3$.*

| T, °C | $t_i$ | $t_e$ |
|---|---|---|
| 300 | 0.82 | 0.18 |
| 400 | 0.78 | 0.22 |
| 500 | 0.76 | 0.24 |
| 600 | 0.72 | 0.28 |
| 700 | 0.71 | 0.29 |

**Table 5.**
*Performances of electronic and ionic conductivity of $(CeO_2)_{0.90}(Sm_2O_3)_{0.10}$ [18].*

conductivity compared with the latter one. The observed plot shapes and level of conductivity are similar to those for lanthanum nickelate and cobaltite with only small differences. However, the addition of iron results in a prominent increase in the transition temperature from semiconductor to metallic conductivity.

Using the West-Tallan method, the ratio of the electronic and ionic conductivity in the studied ceramic samples was determined. As an example, **Table 5** presents the data on the ratio of the transfer numbers of ions and electrons for the studied samples of the composition $(CeO_2)_{0.90}(Sm_2O_3)_{0.10}$. These data indicate that these solid electrolytes have mixed conductivity with the ion transport number—$t_i$ = 0.82 at 300°C and 0.71 at 700°C. The temperature growth leads to a sharp increase in the contribution of the electronic component to the total value of electrical conductivity that relates to a partial transition $Ce^{4+} \rightarrow Ce^{3+}$.

The ratio between the electronic and ionic conductivity determined according to the West-Tallan method is exemplarily illustrated in **Table 6**, indicating the ratios of ion and electron transfer numbers for $La_{0.6}Sr_{0.4}CoO_3$, $La_{0.7}Sr_{0.3}CoO_3$, $La_{0.6}Sr_{0.4}NiO_3$, and $La_{0.7}Sr_{0.3}NiO_3$. The presented data show that these materials have mixed conductivity with a predominance of electronic components with the transfer numbers $t_e$ = 0.92–0.98 and $t_i$ = 0.08–0.02 at 800°C. The electronic component contribution to the total electrical conductivity sharply grows with temperature due to the appearance of metallic conductivity.

| **Composition** | $La_{0.6}Sr_{0.4}CoO_3$ | | $La_{0.7}Sr_{0.3}CoO_3$ | | $La_{0.6}Sr_{0.4}NiO_3$ | | $La_{0.7}Sr_{0.3}NiO_3$ | |
|---|---|---|---|---|---|---|---|---|
| **T, °C** | $t_i$ | $t_e$ | $t_i$ | $t_e$ | $t_i$ | $t_e$ | $t_i$ | $t_e$ |
| 500 | 0.14 | 0.86 | 0.16 | 0.84 | 0.13 | 0.87 | 0.10 | 0.90 |
| 600 | 0.11 | 0.89 | 0.13 | 0.87 | 0.10 | 0.90 | 0.07 | 0.93 |
| 700 | 0.08 | 0.92 | 0.09 | 0.91 | 0.07 | 0.93 | 0.03 | 0.97 |
| 800 | 0.04 | 0.96 | 0.05 | 0.95 | 0.02 | 0.98 | 0.02 | 0.98 |

| **Composition** | $La_{0.6}Sr_{0.4}Fe_{0.7}Ni_{0.3}O_3$ | | $La_{0.7}Sr_{0.3}Fe_{0.7}Ni_{0.3}O_3$ | |
|---|---|---|---|---|
| **T, °C** | $t_i$ | $t_e$ | $t_i$ | $t_e$ |
| 500 | 0.22 | 0.78 | 0.25 | 0.75 |
| 600 | 0.18 | 0.82 | 0.20 | 0.80 |
| 700 | 0.13 | 0.87 | 0.15 | 0.85 |
| 800 | 0.06 | 0.94 | 0.08 | 0.92 |

**Table 6.**
*Performances of electronic and ionic conductivity of* $La_{0.6}Sr_{0.4}CoO_3$, $La_{0.7}Sr_{0.3}CoO_3$, $La_{0.6}Sr_{0.4}NiO_3$, $La_{0.7}Sr_{0.3}NiO_3$, $La_{0.6}Sr_{0.4}Fe_{0.7}Ni_{0.3}O_3$, *and* $La_{0.7}Sr_{0.3}Fe_{0.7}Ni_{0.3}O_3$ *ceramics.*

## 3. Conclusions

Finely dispersed xerogel nanopowders of the compositions $(CeO_2)_{1-x}(Sm_2O_3)_x$ (x = 0.02, 0.05, 0.10), $La_{1-x}Sr_xCo(Ni)O_3$, and $La_{1-x}Sr_xFe_{0.7}Ni_{0.3}O_3$ (x = 0.3, 0.4) with an average crystallite size of ~8–10 nm are obtained by co-crystallization of salts with ultrasonic treatment. Their consolidation by cold uniaxial pressing at the pressure of 150 or 100 MPa followed by sintering at 900–1300°C provided ceramic electrolyte and cathode materials. At 600–1300°C, the obtained ceramic materials are single-phase solid solutions with a fluorite type cubic structure for the $CeO_2$–$Sm_2O_3$ system, perovskite-type tetragonal, and orthorhombic structure for $La_2O_3$–SrO–$Co(Ni)_2O_3$ and $La_2O_3$–SrO–$Fe_2O_3$-$Ni_2O_3$ systems. The obtained ceramic materials are characterized by CSR 64–81 nm (1300°C), open porosity in the range of 2–30%, and relative density of 97–94%. Materials based on cerium oxide have the conductivity $\sigma_{700°C} = 1.3{\cdot}10^{-2}$ S/cm of predominantly ionic type ion transfer numbers $t_i$ = 0.82–0.71 in the temperature range of 300–700°C, due to the formation of mobile oxygen vacancies during the heterovalent substitution of $Ce^{4+}$ for $Sm^{3+}$. Solid solutions based on nickelate and lanthanum cobaltite have mixed electron-ionic conductivity $\sigma_{700°C} = 0.80{\cdot}10^{-1}$ S/cm with transfer numbers $t_e$ = 0.98–0.90, $t_i$ = 0.02–0.10. Lanthanum nickelate features a higher electrical conductivity compared to lanthanum cobaltite; both solid solutions are characterized by an increase in electrical conductivity with the content of strontium oxide. In addition, ceramics with a perovskite-type tetragonal crystal structure show higher electrical conductivity compared to materials with an orthorhombic perovskite-type crystal structure.

The commensurability of TEC values of the resulting electrolyte material based on cerium dioxide ($12.2–12.5{\cdot}10^{-6}\ K^{-1}$) and cathode materials based on lanthanum nickelate ($12.8–14.2{\cdot}10^{-6}\ K^{-1}$) has been established, which makes it possible to consider this pair of ceramic materials as SOFC component.

The obtained ceramics materials according to their mechanical (open porosity, density and thermal expansion coefficient) and electrophysical properties are promising as solid oxide electrolytes and cathodes of medium-temperature fuel cells.

## Author details

Marina V. Kalinina[1*], Daria A. Dyuskina[1], Irina G. Polyakova[1], Sergey V. Mjakin[2], Maxim Yu. Arsent'ev[1] and Olga A. Shilova[1,2,3]

1 Institute of Silicate Chemistry of the Russian Academy of Sciences, St. Petersburg, Russia

2 Saint-Petersburg State Institute of Technology (Technical University), St. Petersburg, Russia

3 Saint-Petersburg State Electrotechnical University "LETI", St. Petersburg, Russia

*Address all correspondence to: tikhonov_p-a@mail.ru

## References

[1] Brusencov VP. Solid Oxide Fuel Cells: Collection of Scientific and Technical Articles [M]. Snezhinsk: Publisher Russian Federal Nuclear Center, All-Russian Research Institute of Technical Physics; 2003. p. 376

[2] Kasyanova AV, Tarutina LN, Rudenko AO, Lyagaeva JG, Medvedev DA. Ba(Ce,Zr)$O_3$-based electrodes for protonic ceramic electrochemical cells: Towards highly compatible functionality and triple-conducting behavior. Russian Chemical Reviews. 2020;**89**(6):667-692. DOI: 10.1070/RCR4928

[3] Pikalova EY, Kalinina EG. Solid oxide fuel cells based on ceramic membranes with mixed conductivity: Improving efficiency. Russian Chemical Reviews. 2021;**90**(6):703-749. DOI: 10.1070/RCR4966

[4] Ponomareva AA, Ivanova AG, Shilova OA, Kruchinina IY. Current state and prospects of manufacturing and operation of methane-based fuel cells (review). Glass Physics and Chemistry. 2016;**42**(1):1-19. DOI: 10.1134/S1087659616010120

[5] Nakamura A, Wagner JB. Defect structure, ionic conductivity, and diffusion in yttria stabilized zirconia and related oxide electrolytes with fluorite structure. Journal of the Electrochemical Society. 1986;**133**(8):1542-1548. DOI: 10.1149/1.2108965

[6] Pal'guev SF, Gil'derman VK, Zemcov VI. High Temperature Oxide Electronic Conductors for Electrochemical Devices. Moskow: Nauka Publ; 1990. p. 198

[7] Sadykov V, Usoltsev V, Yeremeev N, Mezentseva N. Functional nanoceramics for intermediate temperature solid oxide fuel cells and oxygen separation membranes. Journal of the European Ceramic Society. 2013;**33**(12):2241-2250. DOI: 10.1016/j.jeurceramsoc.2013.01.007

[8] Hendriksen PV, Larsen PH, Mogensen M, Poulsen FW. Prospects and problems of dense oxygen permeable membranes. Catalysis Today. 2000;**56**:283-295. DOI: 10.1016/S0920-5861(99)00286-2

[9] Bragg L, Claringbull GF. Crystal Structures of Minerals. London: G. Bell and sons LTD; 1965. p. 409

[10] Jun A, Kim J, Shin J, Kim G. Perovskite as a cathode material: A review of its role in solid-oxide fuel cell technology. ChemElectroChem. 2016;**3**: 511-530. DOI: 10.1002/celc.201500382

[11] Sadykov VA, Pavlova SN, Kharlamova TS, Muzykantov VS. Perovskites and their nanocomposites with fluorite-like oxides as materials for solid oxide fuel cells cathodes and oxygen-conducting membranes: Mobility and reactivity of the surface/bulk oxygen as a key factor of their performance. Perovskites: Structure, Properties and Uses, Nova Science Publishers Inc. USA. 2010;**42**(37):67-178. DOI: 10.1002/chin.201137217

[12] Prasad DH, Son J-W, Kim B-K, Lee H-W, Lee J-H. Synthesis of nano-crystalline $Ce_{0.9}Gd_{0.1}O_{1.95}$ electrolyte by novel sol–gel thermolysis process for IT-SOFCs. Journal of the European Ceramic Society. 2008;**28**:3107-3112. DOI: 10.1016/j.jeurceramsoc.2008.05.021

[13] Kovalenko AS, Shilova OA, Morozova LV, Kalinina MV, Drozdova IA, Arsent'Ev MY. Features of

the synthesis and the study of nanocrystalline cobalt-nickel spinel. Glass Physics and Chemistry. 2014; **40**(1):106-113. DOI: 10.1134/S1087659614010131

[14] Duran P, Villegas M, Capel F, Recio P, Moure C. Low temperature sintering and microstructural development of nano scale Y-TZP ceramics. Journal of the European Ceramic Society. 1996;**16**:945. DOI: 10.1016/0955-2219(96)00015-5

[15] Russian Standard GOST 473.4-81. 1981. (in Russian)

[16] Shilova OA, Antipov VN, Tikhonov PA, Kruchinina IY. Ceramic nanocomposites based on oxides of transition metals of ionistors. Glass Physics and Chemistry. 2013;**39**:570-578. DOI: 10.1134/S1087659613050179

[17] Pivovarova AP, Strakhov VI, Popov VP. On the mechanism of electron conductivity in lanthanum metaniobate. Technical Physics Letters. 2002;**28**(10): 815-817. DOI: 10.1134/1.1519016

[18] Kalinina MV, Dyuskina DA, Fedorenko NY, Shilova OA. Effect of liquid-phase synthesis method of nanopowders on microstructure and physics and chemical properties of ceramics in $CeO_2$-$Sm_2O_3$ system. Inorganic Materials: Applied Research. 2021;7:66-75. DOI: 10.30791/1028-978X-2021-7-66-75

[19] Morozova LV, Kalinina MV, Arsent'ev MY, Popov VP, Drozdova IA, Polyakova IG, et al. Neodymium nickelate – A cathode material for fuel cells. Glass Physics and Chemistry. 2016; **42**(1):95-99. DOI: 10.1134/S108765961 6010089

[20] Yaroslavtsev IY, Bogdanovich NM, Vdovin GK, Dem'Yanenko TA, Bronin DI, Isupova LA. Cathodes based on rare-earth metal nickelate ferrites prepared from industrial raw materials for solid oxide fuel cells. Russian Journal of Electrochemistry. 2014;**50**(6):548-553. DOI: 10.1134/S1023193514060123

[21] Kol'chugin AA, Pikalova EY, Bogdanovich NM, Bronin DI, Filonova EA. Electrochemical properties of doped lantanum–nickelate-based electrodes. Russian Journal of Electrochemistry. 2017;**53**(8):826-833. DOI: 10.1134/S1023193517080110

[22] Kakinuma K, Arisaka T, Yamamura H. Thermal expansion and electrical conductivity of perovskite oxide $(Ln_{1-x}Sr_x)CoO_{3-\delta}$ (ln=La, Nd and Sm). Journal of the Ceramic Society of Japan. 2004;**112**:342-346. DOI: 10.2109/jcersj.112.342

[23] Kroger FA. The Chemistry of Imperfect Crystals. Amsterdam: North-Holland; 1964. p. 1039. DOI: 10.1126/SCIENCE.145.3627.40